Electric Discharge Machining of Insulating and Weakly Conductive Engineering Ceramics

Yonghong Liu　Renjie Ji　Baoping Cai
Yanzhen Zhang　Xiaopeng Li

Responsible Editor: Qunxia Wan   Juan Luo

图书在版编目(CIP)数据

绝缘及弱导电工程陶瓷电火花加工 = Electric Discharge Machining of Insulating and Weakly Conductive Engineering Ceramics：英文 / 刘永红等著. —北京：科学出版社, 2015. 6
Ⅰ. ①绝⋯ Ⅱ. ①刘⋯ Ⅲ. ①电工陶瓷–电火花加工–英文 Ⅳ. ①TM28
中国版本图书馆 CIP 数据核字(2015)第 062367 号

Published by Science Press
16 Donghuangchenggen North Street
Beijing 100717, P. R. China
Printed in Beijing

Copyright© 2015 by Science Press
ISBN 978-7-03-044004-4

All rights reserved. No part of this publication may be reproduced, stored in a retrieval system, or transmitted in any form or by any means, electronic, mechanical, photocopying, recording or otherwise, without the prior written permission of the copyright owner.

Foreword

China officially initiated 211 Project approved by the State Council in 1995. It was the biggest and most important project that was authorized by China in the higher education since the founding of New China. As a response to the domestic and international situation at the turn of the century, Chinese leaders made the important decision to develop the higher education 211 Project drives the overall development of participating universities through major innovations. It encompasses developing university subjects and teaching faculties, as well as other core elements aimed at improving university standards. The 211 project thereby explores successful ways to develop the high level universities.211 Project has been remarkably successful over the last 17 years. It has improved the overall educational quality of chinese higher level education, scientific research standards, and its institutional management and administration. The project has established the foundation for china to help top ranking universities operate at an advanced global level.

In 1997, the China University of Petroleum(UPC) was included in the 211 Project rankings, providing us with an opportunity to develop into a high level university. During the three phases (during the Ninth, Tenth and Eleventh Five-Year Plan periods) of 211 Project to date, the UPC has focused on improving the university's level; stated our mission to meet the needs of the petroleum and petrochemical industry; stated our goal to realize major innovative breakthroughs for national oil and gas; and aimed to improve of key discipline levels, create academic leaders, and cultivate international and innovative talents. We has adhered to the 211 Project development guidelines, and has used our advantages to drive overall improvements and developments. Our competitiveness has been significantly strengthened, and the university's administration and overall strength have noticeably improved, establishing a solid foundation for a world-class petroleum research university.

Participation in 211 Project has strengthened the university's petroleum characteristics and has highlighted our academic advantages. Moreover, we are smoothly implementing our specialty innovation platform. Five of the UPC's national key disciplines and two of our state key (cultivation) disciplines are at leading domestic/international advanced levels.

The UPC's engineering and the chemistry departments entered the Essential Science Indicators' world rankings for the first time in March 2012, indicating that the strength and the level of the two main subjects (the petroleum and petrochemical disciplines) have significantly increased. Our high-level teaching staff structure has substantially progressed. Our staff includes members of the Chinese Academy of Science and Engineering, distinguished Changjiang Scholar professors, national science fund for distinguished young scholars winners, and national 'Thousand Person Plan'and 'Millions of Talents' project candidates among other high-level talents. This provides an intellectual guarantee for the university's future. Innovation ability has substantially improved, and high-level programs and achievements are constantly emerging. Our annual scientific research funds are over 400 million RMB. The UPC has preliminarily established a science and technology innovation system with distinctive petroleum features, and has become an important element of the national science and technology innovation system. Our ability to cultivate innovative talents is improving. The UPC has undertaken a plan to educate and train outstanding engineers, and established a special zone for outstanding innovative talents. We are actively exploring ways to cultivate international talents, and have reformed our postgraduate education mechanism, preliminarily formed an innovative talents cultivation method and a postgraduate education mechanism corresponding to innovative talent cultivation.

We are perfecting our public service support system, We has developed an efficient and fast public service system, significantly upgraded our software and hardware.This provides a strong support basis to develop teaching, scientific research and management levels.

The experiences gained in the 17 years that UPC has been involved in 211 Project are invaluable.First, we must adhere to our guidelines and pioneer discipline innovations by strengthening our characteristics and stressing our advantages. We can thereby realize the goal of building a world-class petroleum department. Second, we must keep abreast of ongoing developments and overall improvements to further widen our advantages. We must coordinate our development and constantly improve our overall competitiveness by promoting the whole through the key parts.Third, we must adhere to the goal of improving our educational mechanisms and establishing our research platform. We must strengthen the overall university structure, collect resources, integrate our talent teams, and optimize all of the management sessions and working systems. We will achieve this by perfecting our educational mechanisms: focusing on elements such as integration, openness, common sharing, competitiveness, information flow, and our project administration mechanisms to ensure that all project work is conducted efficiently and effectively. Fourth, we must focus on staff recruitment and development, and cultivate a talented and united workforce. We must attract outstanding talents to form a team with dedicated and innovative members to promote 211 Project development, which support all of the university's undertakings. Internal schools, disciplines and relevant departments should coordinate and cooperate with one another to form a strong joint effort and guarantee that all of the development goals are smoothly implemented.

We gained invaluable experience since the implementation of 211 Project. It is necessary to carry these experiences forward to further develop the university into an outstanding research institute. To summarize the university's 211 Project experience and its successful outcomes, the UPC initiated a special fund in 2008 to finance the publication of a series of academic monographs relating to the project. These publications relate to both the implementation of 211 Project in the UPC and the scientific research achievements of talented UPC scholars. These monographs introduce and demonstrate the subject construction, scientific innovation and talent cultivation in different categories.

I believe that this series of monographs will disseminate our advanced scientific research results and our academic thoughts from different perspectives and multiple dimensions, and demonstrate the remarkable achievements made and development route taken during our implementation of 211 Project to date. They will support our social influence, improve our academic reputation and make an important and unique contribution to our ongoing 211 Project development plans.

Finally, I am grateful to all of our scholars for their hard work and considerable efforts for 211 Project. I commend the monograph authors in the dedication of collecting and summarizing all our research results. They will leave innovative results and academic spirits, and make invaluable contributions to our university, our faculties, the staffs and the students.

President of China University of Petroleum (East China)
September 2012

Preface

Engineering ceramics have the advantages of good mechanical properties, chemical and thermal stabilities at elevated temperature and excellent electrical performance. Hence, they have been widely used recent years in the machinery, electronics, metallurgy, chemical, geology, aerospace, and nuclear industries. The requirement on the precision, efficiency, and surface integrity of the component is higher and higher. However, it shows low efficiency, high cost, and poor surface integrity when machining them with the present machining methods. Electric discharge machining (EDM) is a non-traditional machining process removing material by a succession of repeated electrical discharges between an electrode and a workpiece. Since the electrode does not contact with the workpiece during machining, EDM is an effective and economical technique for machining conducting ceramics with any rigidity. However, the usually engineering ceramics are insulating or weakly conductive, and EDM cannot be directly used to machine these materials. The development of the process for electric discharge machining of insulating and weakly conductive engineering ceramics is an important research topic in the field of non-traditional machining. The authors have been engaged in the theoretical analysis and experimental investigation on this topic for more than twenty years, and obtained some achievements.

Therefore, we write the book, focusing on the various new machining methods and the machining mechanism for electric discharge machining of insulating and weakly conductive engineering ceramics. More people can understand the progress about the topic by this book, by which to promote the application of the insulating and weakly conductive engineering ceramics, enlarge the application field of electrical discharge machining technology, and enrich the theory of non-traditional machining.

The book is structured as follows:

Chapter 1 presents a new process of machining insulating engineering ceramics using electrical discharge (ED) milling, which is able to effectively machine a large surface area on insulating ceramics, and effectively machine other advanced insulating materials such as cubic boron nitride (CBN), polycrystalline diamond (PCD). The machining principle and characteristics of the technique are introduced. The effects of various machining parameters on the process performance have been investigated.

Moreover, the mathematical models of thermal eroding insulating ceramics with a single pulse discharge are established.

Chapter 2 presents a new method which employs a high energy capacitor for electric discharge machining of insulating ceramics efficiently. The process uses the high voltage, large capacitor and high discharge energy, it is able to effectively machine insulating ceramics, and the single discharge crater volume of insulating ceramics can reach 17.63mm^3. The effects of polarity, peak voltage, capacitance, current-limiting resistance, tool electrode feed, tool electrode section area and assisting electrode thickness on the process performance such as the single discharge crater volume, the tool wear ratio and the assisting electrode wear ratio have been investigated.

Chapter 3 focuses on electrical discharge and grinding with synchronous servo double electrodes (EDGSSDE) for insulating engineering ceramics which integrates electrical discharge and mechanical grinding with a metal bonded diamond conductive wheel. The effects of electrical parameters on material removal rate and surface roughness of the insulating ceramics are discussed. Moreover, a finite element-based analysis of residual stresses in insulating ceramics with this method is presented.

Chapter 4 investigates the effects of the electrical resistivity and the EDM parameters on the EDM performance of engineering ceramics in terms of the machining efficiency and the quality. Moreover, this chapter employs a steel toothed wheel as the tool electrode to machine SiC ceramics with specific resistivity of 500Ω·cm using electrical discharge milling, which is able to effectively machine a large surface area on SiC ceramics with electrical resistivity of 500Ω·cm, and effectively machine other advanced materials with high electrical resistivity such as polycrystalline diamond, and cubic boron nitride.

Chapter 5 presents a new process of machining weakly conductive engineering ceramics using end electrical discharge milling, which employs a turntable with several small cylindrical rods as the tool electrode, and uses a water-based emulsion as the machining fluid. The machining principle and characteristics of the technique are introduced. The effects of machining parameters on the process performance have been investigated with Taguchi experimental design method. Analysis of variance (ANOVA) and *F*-test are used to indicate the significant machining parameters affecting the machining characteristics. Furthermore, mathematical models relating to the machining characteristics are established with the stepwise regression method.

Chapter 6 proposes a new technology of machining SiC ceramics with electrical discharge milling and mechanical grinding compound method. The effects of pulse duration, pulse interval, peak voltage, peak current and feed rate of the workpiece on the process performance such as material removal rate, relative electrode wear ratio and surface roughness have been investigated. A L_{25} orthogonal array based on Taguchi method is adopted, and the experimental data are statistically evaluated by analysis of variance and stepwise regression. In addition, the surface microstructures machined by the new process have been observed by a scanning electron microscope (SEM), an X-ray diffraction (XRD) and an energy dispersive spectrometer (EDS).

Chapter 7 presents a compound process that integrates end electrical discharge milling and mechanical grinding to machine weakly conductive SiC ceramics. The process employs a turntable with several uniformly-distributed cylindrical copper electrodes and abrasive sticks as the tool, and uses a water-based emulsion as the machining fluid. End electrical discharge milling and mechanical grinding happen alternately, and are mutually beneficial, so the process is able to effectively machine a large surface area on SiC ceramic with a good surface quality. The machining principle and characteristics of the technique are introduced. Moreover, Taguchi experimental design, analysis of variance and stepwise regression are used to investigate the effects of machining parameters on performance parameters and surface integrity.

Chapter 8 fouses on the machining fluid for electrical discharge machining of engineering ceramics. Three kinds of emulsion are developed as the dielectric during electrical discharge machining of engineering ceramics, and the effects of dielectric and machining parameters on the process performance have been investigated.

In writing this book, we have benefited a great deal from a large number of individuals who have contributed immensely to this work through either valuable discussions, comments on earlier drafts, or strongly believing in this project and how it was conducted. In this regard, we are indebted to Jianhua Zhang, Yanting Zhang, Guoming Chen, Wensheng Xiao, Qingyun Li, Xiaopeng Li, Lili Yu, Fei Wang, Zengkai Liu, Ruiqiang Diao, Chenchen Xu, and Guangxu Wang. We would also like to thank China University of Petroleum for their support for publishing this book.

Finally, we wish to acknowledge the support of National Natural Science Foundation of China (No. 51275529), National Natural Science Foundation of China (No. 51205411), Taishan Scholar project of Shandong Province (TS20110823), and Shandong Provincial Natural Science Foundation of China (Grant No. ZR2012EEL15).

Prof. Yonghong Liu & Dr. Renjie Ji
Qingdao, China
September, 2014

Contents

Foreword

Preface

Chapter 1 Electric Discharge Milling of Insulating Engineering Ceramics ········· 1
- 1.1 Introduction ·· 2
- 1.2 Principle and characteristics of ED milling ··· 3
  - 1.2.1 Principle of ED milling ·· 3
  - 1.2.2 Characteristics of ED milling ·· 5
- 1.3 Experiments and discussion ··· 5
  - 1.3.1 Effects of the pulse duration on the process performance ················· 6
  - 1.3.2 Effects of the pulse interval on the process performance ·················· 7
  - 1.3.3 Effects of the peak current on the process performance ···················· 8
  - 1.3.4 Effect of tool polarity on the process performance ···························· 9
  - 1.3.5 Effect of peak voltage on the process performance ························· 11
  - 1.3.6 Effect of rotational speed of the tool electrode on the process performance ···· 12
  - 1.3.7 Effect of feed speed of the workpiece on the process performance ············ 13
  - 1.3.8 Effect of emulsion concentration of the machining fluid on the process performance ········· 14
  - 1.3.9 Effect of $NaNO_3$ concentration of the machining fluid on the process performance ········· 15
  - 1.3.10 Effect of Polyvinyl alcohol concentration on the process performance ····· 17
  - 1.3.11 Effect of flow velocity of the machining fluid on the process performance ···· 19
- 1.4 Numerical simulation of single pulse discharge machining insulating Al_2O_3 ceramic ········· 20
  - 1.4.1 Model details ··· 20
  - 1.4.2 Finite element formulations ·· 22
  - 1.4.3 Results and discussion ·· 23
- 1.5 Conclusions ·· 29
- References ·· 30

Chapter 2 Single Discharge Machining of Insulating Ceramics Efficiently with High Energy Capacitor ········· 33
- 2.1 Introduction ·· 33

2.2 Experiments ·· 35
    2.2.1 Experimental principle ·· 35
    2.2.2 Experimental procedure ··· 36
2.3 Results and discussion ··· 37
    2.3.1 Effect of tool polarity on the process performance ············· 37
    2.3.2 Effect of peak voltage on the process performance ············· 39
    2.3.3 Effect of capacitance on the process performance ·············· 40
    2.3.4 Effect of current-limiting resistance on the process performance ··········· 41
    2.3.5 Effect of tool electrode feed on the process performance ··············· 41
    2.3.6 Effect of tool electrode section area on the process performance ············· 43
    2.3.7 Effect of assisting electrode thickness on the process performance ············ 44
2.4 Microstructure character of single discharge crater on insulating ceramic surface ········· 45
2.5 Conclusions ·· 48
References ·· 49

Chapter 3 Electrical Discharge Mechanical Grinding of Insulating Engineering Ceramics ·········· 51

3.1 Introduction ·· 52
3.2 Principle of EDGSSDE ·· 54
3.3 Formation and function of oxide layer on grinding wheel ············· 54
3.4 Formation and function of metamorphosed layer on workpiece surface ··········· 55
3.5 Analysis of residual stresses in EDGSSDE of engineering ceramics ··········· 57
    3.5.1 Assumptions ··· 57
    3.5.2 Temperature models ·· 57
    3.5.3 Residual stresses model ·· 60
    3.5.4 Results ·· 61
3.6 Experiments ·· 63
    3.6.1 Experimental Conditions ·· 63
    3.6.2 Results and analyses ··· 64
3.7 Conclusions ·· 67
References ·· 68

Chapter 4 Electrical Discharge Milling of Weakly Conductive Engineering Ceramics ·········· 70

4.1 Introduction ·· 71

4.2 Experimental procedures for EDM performance of engineering ceramics with different electrical resistivities ... 72
4.3 Results and discussion of EDM performance of engineering ceramics with different electrical resistivities ... 74
    4.3.1 Effect of the electrical resistivity and pulse duration on the process performance ... 74
    4.3.2 Effect of the electrical resistivity and pulse interval on the process performance ... 77
    4.3.3 Effect of the electrical resistivity and peak current on the process performance ... 80
    4.3.4 Microstructure character of ZnO/Al_2O_3 ceramic surface machined by EDM ... 82
4.4 Principle for electric discharge milling of weakly conductive SiC ceramic ... 83
4.5 Experiments and discussion for ED milling of weakly conductive SiC ceramic ... 85
    4.5.1 Effect of tool polarity on the process performance ... 85
    4.5.2 Effect of pulse duration on the process performance ... 87
    4.5.3 Effect of pulse interval on the process performance ... 88
    4.5.4 Effect of peak voltage on the process performance ... 89
    4.5.5 Effect of peak current on the process performance ... 90
    4.5.6 Effect of emulsion concentration on the process performance ... 92
    4.5.7 Effect of milling depth on the process performance ... 93
    4.5.8 Effect of rotational speed on the process performance ... 94
    4.5.9 Effect of tooth number on the process performance ... 95
    4.5.10 Effect of tooth width on the process performance ... 96
    4.5.11 Microstructure character of SiC surface machined by ED milling ... 98
4.6 Conclusions ... 99
References ... 99

Chapter 5 End Electric Discharge Milling of Weakly Conductive Engineering Ceramics ... 102
5.1 Introduction ... 102
5.2 Principle and characteristics for end ED milling of weakly conductive SiC ceramic ... 104
    5.2.1 Principle for end ED milling of SiC ceramic ... 104
    5.2.2 Characteristics for end ED milling of SiC ceramic ... 105
5.3 Experiments ... 107
    5.3.1 Experimental procedures ... 107

		5.3.2	Experimental design	107
		5.3.3	Analysis and discussion of experimental results	108
	5.4	Results and discussion of the single factor experiment during end ED milling		109
		5.4.1	Effect of tool polarity on the process performance	109
		5.4.2	Effect of pulse duration on the process performance	110
		5.4.3	Effect of pulse interval on the process performance	111
		5.4.4	Effect of peak voltage on the process performance	113
		5.4.5	Effect of peak current on the process performance	114
		5.4.6	Effect of emulsion concentration on the process performance	115
		5.4.7	Effect of emulsion flux on the process performance	116
		5.4.8	Effect of milling depth on the process performance	118
		5.4.9	Effect of electrode number on the process performance	119
		5.4.10	Effect of electrode diameter on the process performance	121
	5.5	Results and discussion of the orthogonal experiment during end ED milling		122
		5.5.1	Analysis of the MRR	123
		5.5.2	Analysis of the EWR	126
		5.5.3	Analysis of the SR	128
		5.5.4	Confirmation experiments	131
	5.6	Analysis of the machined surface by end ED milling		131
		5.6.1	SEM observation of the machined surface	131
		5.6.2	Compositions of the machined surface	133
	5.7	Conclusions		137
	References			138
Chapter 6	Electric Discharge Milling and Mechanical Grinding Compound Machining of Weakly Conductive Engineering Ceramics			141
	6.1	Introduction		141
	6.2	Principle for ED milling and mechanical grinding of SiC ceramic		142
	6.3	Experiments		144
		6.3.1	Experimental procedures	144
		6.3.2	Experimental design	146
	6.4	Experimental results and discussion of orthogonal array		147
		6.4.1	Analysis of MRR	148
		6.4.2	Analysis of REWR	151
		6.4.3	Analysis of the SR	154

		6.4.4 Experiment validation ··· 156
6.5		Surface integrity of SiC ceramic machined by the compound process ··· 158
	6.5.1	Surface topography of the machined workpiece ··· 158
	6.5.2	Compositions of the machined workpiece ··· 160
6.6		Conclusions ··· 164

References ··· 165

Chapter 7 High Speed End Electrical Discharge Milling and Mechanical Grinding of Weakly Conductive Engineering Ceramics ··· 168

7.1		Introduction ··· 168
7.2		Principle and characteristics for end ED milling and mechanical grinding of SiC ceramic ··· 170
	7.2.1	Principles ··· 170
	7.2.2	Characteristics of end ED milling and mechanical grinding of SiC ceramic ··· 173
7.3		Experiments ··· 174
	7.3.1	Experimental procedures ··· 174
	7.3.2	Experimental design ··· 175
	7.3.3	Analysis and discussion of experimental results ··· 175
7.4		Results and discussion of the single factor experiment ··· 176
	7.4.1	Effect of tool polarity on the process performance ··· 176
	7.4.2	Effect of pulse duration on the process performance ··· 177
	7.4.3	Effect of pulse interval on the process performance ··· 179
	7.4.4	Effect of open-circuit voltage on the process performance ··· 180
	7.4.5	Effect of discharge current on the process performance ··· 181
	7.4.6	Effect of diamond grit size on the process performance ··· 182
	7.4.7	Effect of emulsion concentration on the process performance ··· 184
	7.4.8	Effect of emulsion flux on the process performance ··· 185
	7.4.9	Effect of milling depth on the process performance ··· 186
	7.4.10	Effect of tool stick number on the process performance ··· 187
7.5		Analysis of Taguchi method ··· 189
	7.5.1	Analysis of Taguchi method for MRR ··· 190
	7.5.2	Analysis of Taguchi method for the TWR ··· 192
	7.5.3	Analysis of Taguchi method for the SR ··· 194
	7.5.4	Confirmation experiments ··· 196
7.6		Surface integrity of SiC ceramic machined by the compound machining ··· 196
	7.6.1	Surface morphology of the machined surface ··· 196
	7.6.2	Surface roughness ··· 198

    7.6.3 Micro-cracks on the machined surface ··· 199
    7.6.4 Compositions of the machined surface ·· 201
  7.7 Conclusions ··· 204
  References ·· 206

Chapter 8 Machining Fluid for Electrical Discharge Machining of Engineering Ceramics ·· 208

  8.1 Introduction ··· 208
  8.2 Experimental procedures ··· 209
  8.3 Effect of the additive on the emulsion property and EDM performance ······ 210
    8.3.1 Effect of the anionic compound emulsifier on the emulsion property ········ 210
    8.3.2 Effect of ACE concentration on EDM performance ···························· 211
    8.3.3 Effect of OP-10 concentration on emulsion property ···························· 213
    8.3.4 Effect of OP-10 concentration on EDM performance ··························· 214
  8.4 Influence of dielectric on the process performance for ED milling of SiC ceramic ··· 216
    8.4.1 Effect of tool polarity on the process performance in different emulsions ···· 218
    8.4.2 Effect of pulse duration on the process performance in different emulsions ···· 221
    8.4.3 Effect of pulse interval on the process performance in different emulsions ··· 222
    8.4.4 Effect of peak voltage on the process performance in different emulsions ··· 224
    8.4.5 Effect of peak current on the process performance in different emulsions ···· 225
    8.4.6 Surface analysis of workpiece ·· 226
  8.5 Conclusions ··· 230
  References ·· 231

Chapter 1　Electric Discharge Milling of Insulating Engineering Ceramics

Engineering ceramics have been used widely in modern industry. However, the manufacture of engineering ceramic blanks is not an efficient process. The shaping of engineering ceramic blanks with conventional machining methods (such as grinding), is a long, labour-intensive and costly process of EDM processes promise to be effective and econmical techniques for the production of tools and parts from ceramic blanks. Wire electric discharge machining (WEDM) is able to effectively slice ceramics. Electrical discharge machining (EDM) and electrical discharge grinding (EDG) shape ceramic blanks at a lower cost. With the help of some assisting methods, WEDM, EDM and EDG can be used to machine insulating ceramics. However, WEDM, EDM and EDG of insulating ceramics show lower efficiency, especially for a large surface area on insulating ceramics. This chapter presents a new process of machining insulating ceramics using ED milling. ED milling uses a thin copper sheet fed to the tool electrode along the surface of workpiece as the assisting electrode, and uses a water-based emulsion as the machining fluid. This process is able to effectively machine a large surface area on insulating ceramics, and effectively machine other advanced insulating matcrials such as cubic boron nitride, polycrystalline diamond. The machining principle and characteristics of the technique are introduced. The effect of pulse duration, pulse interval, peak current, tool polarity, peak voltage, rotational speed of tool electrode and feed speed of workpiece, emulsion concentration, $NaNO_3$ concentration, polyvinyl alcohol concentration and flow velocity of the machining fluid on the process performance has been investigated. Moreover, the mathematical models of thermal eroding insulating Al_2O_3 ceramic with a single pulse discharge are established. The thermal erosion characteristics of insulating Al_2O_3 ceramic with a single pulse discharge have been numerically simulated. The numerical simulation results have been validated by experiment.

1.1 Introduction

The demands for advanced ceramics are continually increasing in industry. Engineering ceramics have been widely used in modern industry such as ballistic armor, ceramic composite automotive brakes, diesel particulate filters, a wide variety of prosthetic products, piezoceramic sensors and next-generation computer memory products, because of their higher hardness and wear resistance, lower thermal expansion coefficient and density, and their chemical inertness [1-3]. However, during sintering processes in the fabrication of ceramic components, volumetric shrinkages of the ceramics inevitably take place, such that machining and surface finishing of the components become necessary to be able to impart an accurate final shape and size to precision machine elements [4,5].

Ceramics are known as very difficult-to-machine materials. The main factors that cause ceramics to be difficult to machine are their high hardness, non-electrical conductivity and brittleness [6]. Diamond grinding is one of the most commonly used techniques for engineering ceramic blank shaping, but it is costly and inefficient [7,8]. The high hardness of ceramics induces higher grinding force and quick wear of diamond cutting edges. The inherent brittleness of ceramics generates micro-cracks in their surfaces and chipping in the corner of ground components, which adversely affect the surface integrity and the quality of the components [9,10].

EDM processes promise to be an effective and economical technique for the production of tool and parts from conducting ceramic blanks. WEDM is able to effectively slice conducting ceramics [11,12]. EDM and EDG shape conducting ceramic blanks at a low cost [13,14]. However, electromachining techniques cannot be directly used to machine insulating ceramics, these materials are non-conducting. The development of the process for EDM insulating ceramics is an important research topic in the field of nontraditional machining, a lot of research achievements have been achieved about the topic.

Some researchers have used electrolyte as the machining fluid to achieve WEDM, EDM, EDG, arc discharge machining, gas-filled electrochemical discharge machining, mechanical electrical discharge, and electrochemical compound machining of insulating ceramics [15-19]. Tani et al. [20] and Fukuzawa et al. [21] developed the techniques of WEDM and EDM insulating ceramics using kerosene as working fluid. In this method, a metal plate or metal mesh is arranged on the surface of insulating ceramics as an assisting electrode. With the help of assisting electrode, insulating ceramics can be machined by sinking EDM or by WEDM in work oil. However, these processes of machining a large surface area on insulating ceramics show low efficiency.

This chapter proposes a new technique of machining insulating ceramics using ED milling. The effect of pulse duration, pulse interval, peak current, tool polarity, peak voltage, rotational speed of tool electrode, and feed speed of workpiece on the process performance has been investigated. Machining fluid is a primary factor that affects the material romoval rate and surface quality for ED milling. The effects of emulsion concentration, $NaNO_3$ concentration, polyvinyl alcohol concentration and flow velocity of the machining fluid on the process performance have been investigated. Moreover, the mathematical models of thermal eroding insulating Al_2O_3 ceramic with a single pulse discharge are established. The thermal erosion characteristics of insulating Al_2O_3 ceramic with a single pulse discharge have been numerically simulated. The numerical simulation results have been validated by experiment.

1.2 Principle and characteristics of ED milling

1.2.1 Principle of ED milling

Insulating ceramics cannot be directly machined by EDM. In order to machine these materials with EDM, conditions for electrical discharge on their surface must to be created. One of the most commonly used methods of generating the discharge conditions is using electrolyte as the machining fluid. However, the process generates harmful gas during machining, it shows low efficiency, and its equipment can be easily eroded by electrolyte. ED milling uses a water-based emulsion as the machining fluid. With the help of assisting electrode, insulating materials as insulating ceramics can be easily machined by ED milling. The schematic illustration of ED milling is shown in Fig. 1.1. The tool electrode and the assisting electrode are connected to the positive and negative poles of the pulse generator respectively. The tool electrode is a steel wheel and is mounted on a rotary spindle, driven by an A.C. (alternating current) motor. The workpiece is insulating ceramic blank and is mounted on an NC (numerically controlled) table. The assisting electrode is a thin copper sheet. The stored sheet bobbin is mounted on to a rotary spindle, driven by a D.C. (direct current) servo motor. The actuating wheel is driven by the D.C. servo motor with a gearing. The machining fluid is a water-based emulsion without electrolysis.

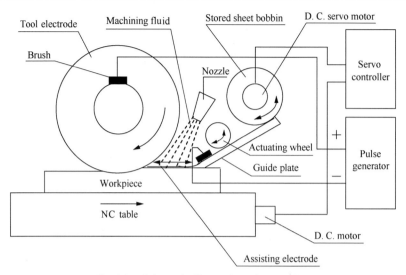

Fig. 1.1 Schematic illustration of ED milling

During machining, the tool electrode rotates at a high speed; the assisting electrode is fed towards the tool electrode along the surface of insulating ceramic workpiece, driven by the actuating wheel. As short-circuits or arcs generated in ED milling, the assisting electrode is fed back by the stored sheet bobbin. After the short circuits or arcs are cleared up, the assisting electrode is fed on again. The machining fluid is flushed to the assisting electrode and the tool electrode with the nozzle. The flushing force of the machining fluid presses the assisting electrode close to the workpiece. As the assisting electrode approaches the tool electrode and the gap between them reaches the discharge gap, electrical discharges are produced. A plasma channel grows during the pulse duration. A vapor bubble forms around this channel. The surrounding water-based emulsion restricts plasma growth, and makes the plasma energy densities to rise very high. Nancy and Ahmed [22] showed that the plasma temperature reached nearly 40000K and plamsa pressure could rise to 300MPa in EDM of conducting advanced cerimics. Comparing with EDM of conducting advanced cerimics, electrical discharge between two metal electrodes is easily produced and water-based emulsion shows a high restricting effect in ED milling, so its plasma temperature can reaches nearly 40000K and plamsa pressure can rise to 300MPa. The high instantaneous temperature and pressure plasma causes insulting ceramics to melt, vapor, and thermally spall. During rough ED milling, most of insulting ceramics removed by thermal spalling. During finish of ED milling, a main factor causing insulting ceramic removal is melting and evaporation. Because the assisting copper sheet electrode is very thin and close to workpiece, most of electrical discharge energy acting on the assisting electrode and plasma energy can be directly acted on the surface of workpiece.

1.2.2 Characteristics of ED milling

The advantages of ED milling as follows.

(1) Using a thin sheet copper close to the surface of insulating ceramics as the assisting electrode, insulating ceramics can be easily machined by ED milling. The process shows high material removal rate (MRR). Fig. 1.2 is the photograph of insulating Al_2O_3 ceramic workpiece machined by ED milling.

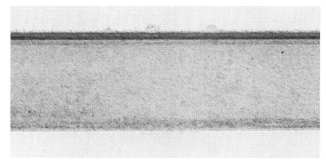

Fig. 1.2　Photograph of insulating Al_2O_3 ceramic workpiece machined by ED milling

(2) Using a water-based emulsion as the machining fluid, harmful gas is not generated during ED milling, and its equipment is not corroded.

(3) Using steel as tool material, the tool electrode is manufactured easily and shows low cost.

(4) Using a conducting abrasive wheel as the tool electrode, the compound machining process with EDM and mechanical grinding can be easily performed.

(5) With a rotary tool electrode, the eroded chips are easy to flush away. Therefore, Ed milling improves the stability of the electrical discharges.

(6) With some specific motion controls, ED milling can machine cylindrical surface, conical surfaces and spherical surfaces.

1.3　Experiments and discussion

In the following experiments, the workpiece material was insulating Al_2O_3 ceramic, the tool electrode material was steel, the assisting electrode material was red copper and the machining fluid was a water-based emulsion.

1.3.1 Effects of the pulse duration on the process performance

Fig. 1.3 shows that the MRR initially increases fast with an increase of pulse duration and then increases slowly with an increase in pulse duration, for pulse interval of 400μs, peak current of 25A, and polarity of tool electrode (+)/assisting electrode (−). There are many reasons causing the phenomena. The longer the pulse duration with less current generated the more thermal energy is lost due to heat conduction; therefore the MRR is low. However, with a very short pulse duration and a high current level, the thermal energy density is very high. At this time, the material is removed by vaporization. As the vaporization heat consumption is high, the MRR is low.

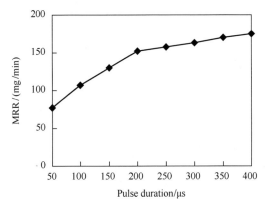

Fig. 1.3　Effect of pulse duration on MRR

Fig. 1.4 shows the effect of pulse duration on surface roughness (SR), for pulse interval of 400μs, peak current of 25A, and polarity of tool electrode (+)/assisting electrode (−). It can be seen from Fig. 1.4 that SR initially increases with an increase in pulse duration, and then decreases slowly with an increase in pulse duration, because the crater size generated by a single pulse becomes large with an increase in single pulse energy. Single pulse energy increases with increasing pulse duration; therefore the SR rises with an increase in pulse duration. As the pulse duration is longer than 200μs, in a single pulse duration several electric discharges are generated, as shown in Fig. 1.5, the single electric discharge energy becomes small; so the SR decreases slowly with an increase in pulse duration.

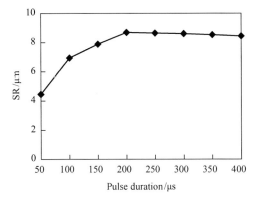

Fig. 1.4 Effect of pulse duration on SR

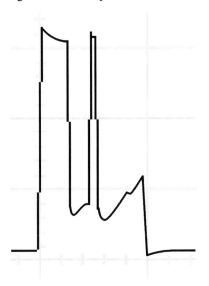

Fig. 1.5 Voltage wave of several electrical discharges in a pulse duration

1.3.2 Effects of the pulse interval on the process performance

Fig. 1.6 shows the effect of pulse interval on the MRR, for pulse duration of 500μs, peak current of 25A, and polarity of tool electrode (+)/assisting electrode (−). The MRR decreases with an increase of pulse interval, the reason is that the electrical discharge frequency decreases with an increase of pulse interval, so the MRR decreases.

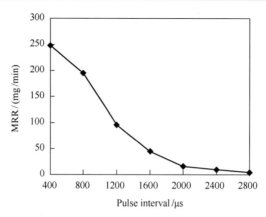

Fig. 1.6　Effect of pulse interval on MRR

As shown in Fig. 1.7, the SR decreases with an increase in pulse interval, for pulse duration of 500μs, peak current of 25A, and polarity of tool electrode (+)/assisting electrode (–). With a longer pulse interval, there is more time to grind the modified layer on the workpiece and clear the disintegrated particles from the gap between the tool electrode and the workpiece, there is also more time for deionization of the dielectric; therefore, the SR decreases with an increase of pulse interval.

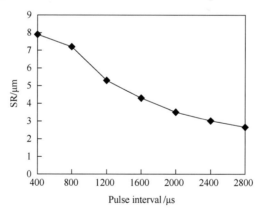

Fig. 1.7　Effect of pulse interval on SR

1.3.3　Effects of the peak current on the process performance

Fig. 1.8 shows the relationship between the MRR and peak current, for pulse duration of 500μs, pulse interval of 400μs, and polarity of tool electrode (+)/assisting electrode (–). The MRR rises with an increase in peak current. The phenomenon can be explained by relationship between the material removed by the single pulse and peak current as follows [23].

$$w_0 = K_{w_0} T_i I_p \qquad (1.1)$$

Where, w_0 is material removal by a single pulse theoretically developed (g/pulse), K_{w_0} is constant, I_p is peak current, T_i is pulse duration.

It can be seen from Equation (1.1) that under a certain pulse duration, the material removed by a single pulse increases with an increase in peak current; therefore, MRR increase. As the peak current is less than 5A, the MRR is less than 0. The reason for this is that the weight of insulating ceramics removed by ED milling is less than the weight of the tool electrode and assisting electrode eroded by ED milling that is plated on the workpiece.

The effect of peak current on SR is shown in Fig. 1.9, for pulse duration of 500μs, pulse interval of 400μs, and polarity of tool electrode (+)/assisting electrode (−). SR initially rises with an increase in peak current, and then decreases with an increase in peak current. The reason for this is that the material removed by a single pulse rises with an increase in peak current; therefore SR increase. As the peak current is larger than 20A, the thickness of modified layer on the workpiece increases, the material removed by the tool electrode grinding increases; the SR is low.

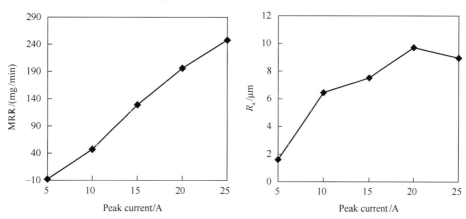

Fig. 1.8 Effect of peak current on MRR   Fig. 1.9 Effect of peak current on SR

1.3.4 Effect of tool polarity on the process performance

In ED milling, the choice of tool electrode polarity is an important factor. The effect of tool polarity on the material removal rate (MRR) and the SR is illustrated respectively in Fig. 1.10 and Fig. 1.11, for pulse duration of 50μs and 500μs, pulse interval of 350μs, and peak current of 25A.

MRR with different tool polarities is given in Fig. 1.10. Under the same conditions, the MRR in positive tool polarity is 2~4 times that with negative tool polarity. This phenomenon can be explained as follows. The longer pulse duration is used in the ED milling, the positive ions of discharge channel have enough time to be accelerated and the numbers of the positive ions arriving at the negative assisting electrode polarity close to the workpiece increase. Because the mass of the positive ions is much larger than that of electrons, the bombardment effect by the positive ions is stronger than that by electrons; therefore, MRR is high in positive tool polarity.

Fig. 1.11 shows the influence of tool polarity on SR. Under the same conditions the SR in positive tool polarity is higher than that with negative tool polarity. This is because the crater size generated by the positive ions is larger than that by electrons; therefore, the SR is high in positive tool polarity.

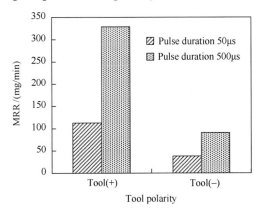

Fig. 1.10   Effect of tool polarity on MRR

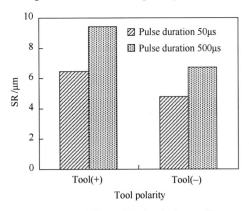

Fig. 1.11   Effect of tool polarity on SR

1.3.5 Effect of peak voltage on the process performance

The effect of peak voltage on the MRR and the SR is illustrated respectively in Fig. 1.12 and Fig. 1.13, for pulse duration of 500μs, pulse interval of 350μs, peak current of 25A, and tool electrode as positive polarity.

Fig. 1.12 shows the relationship between the MRR and peak voltage. The MRR rises with an increase in peak voltage. The phenomenon can be explained by relationship between the material removed by the single pulse and peak voltage as follows [23].

$$w_0 = K_{w_0} T_i \frac{U_p}{R} \quad (1.2)$$

Where, w_0 is material removal by a single pulse theoretically developed (g/pulse), K_{w_0} is constant, T_i is pulse duration, U_p is peak voltage, R is current-limit resistance.

It can be seen from Equation (1.2) that under the pulse duration and current-limit resistance are constant, the material removed by a single pulse increases with an increase in peak voltage; therefore MRR increases.

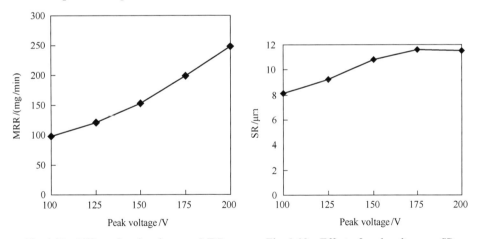

Fig. 1.12 Effect of peak voltage on MRR   Fig. 1.13 Effect of peak voltage on SR

The effect of peak voltage on SR is shown in Fig. 1.13. The SR initially rises with an increase in peak voltage, and then decreases with an increase in peak voltage. The reason for this is that the material removed by a single pulse rises with an increase in peak voltage; therefore, SR increase. As the peak voltage is higher than 175V, the thickness of modified layer on the workpiece increases, the material removed by the tool electrode milling increases; the SR is low.

1.3.6 Effect of rotational speed of the tool electrode on the process performance

The effect of rotational speed of the tool electrode on the MRR and the SR is illustrated respectively in Fig. 1.14 and Fig. 1.15, for pulse duration of 500μs, pulse interval of 350μs, peak current of 25A, and tool electrode as positive polarity.

As shown in Fig. 1.14, the MRR initially increases fast with an increase in rotational speed of the tool electrode, and then increases slowly with an increase in rotational speed of the tool electrode. There are many reasons causing the phenomena. The velocity of mechanical milling workpiece by tool electrode increases with an increase in rotational speed of the tool electrode. The high rotational speed of the tool electrode enhances the flow velocity of the work fluid flushing to the discharge gap and the assisting electrode, it makes the assisting electrode becomes more close to the surface of the workpiece. The stability and electrical discharge energy acting on the surface of workpiece increases, the material removed by electrical discharge rises. Therefore, the MRR increases with an increase in rotational speed of the tool electrode.

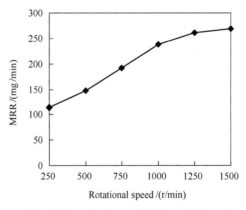

Fig. 1.14　Effect of rotational speed of the tool electrode on MRR

Fig. 1.15 shows the influence of rotational speed of the tool electrode on SR. The SR initially increases with an increase in rotational speed of the tool electrode, and then increases slowly with an increase in rotational speed of the tool electrode. Because the material removed by electric discharge becomes strong with an increase in rotational speed of the tool electrode, the crater size generated by electric discharge becomes large, the SR rises. As the rotational speed of the tool electrode is higher than a certain value, the material removed by mechanical milling rises fast, and makes the crater size generated by electric discharge becomes small; the SR rises slowly.

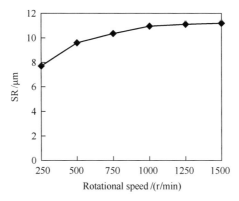

Fig. 1.15 Effect of rotational speed of the tool electrode on SR

1.3.7 Effect of feed speed of the workpiece on the process performance

The effect of feed speed of the workpiece on the MRR and the SR is illustrated respectively in Fig. 1.16 and Fig. 1.17, for pulse duration of 500μs, pulse interval of 350μs, peak current of 25A, and tool electrode as positive polarity.

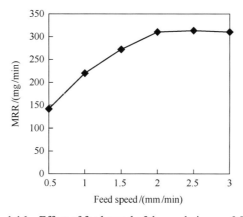

Fig. 1.16 Effect of feed speed of the workpiece on MRR

As shown in Fig. 1.16, the MRR initially increases fast with an increase in feed speed of the workpiece, and then decreases with an increase in feed speed of the workpiece. The reason for this is that under the same milling width and depth of the tool electrode, the material removed by ED milling increases with an increase in feed speed of the workpiece. As the feed speed of the workpiece is higher than the maximum machining speed by ED milling, mechanical vibration of the tool electrode becomes strong, short-circuits or arcs are easily generated in DE milling; the MRR is low.

The effect of feed speed of the workpiece on SR is shown in Fig. 1.17. The SR initially decreases with an increase in feed speed of the workpiece, and then increases slowly with an increase in feed speed of the workpiece. The time of repeat electrical discharge at a certain place decreases at the fast feed speed of the workpiece; therefore, the SR is low. As the feed speed of the workpiece is higher than the maximum machining speed by ED milling, mechanical vibration of the tool electrode becomes stronger, short-circuits or arcs are easily generated in DE milling; the SR rises.

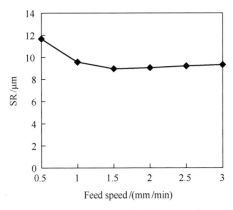

Fig. 1.17　Effect of feed speed of the workpiece on SR

1.3.8　Effect of emulsion concentration of the machining fluid on the process performance

The effect of emulsion concentration on the MRR and the SR is illustrated respectively in Fig. 1.18 and Fig. 1.19; the machining fluid was water and emulsion.

As shown in Fig. 1.18, the MRR initially increases fast with an increase in emulsion concentration, and then decreases slowly with an increase in emulsion concentration. There are many reasons causing the phenomena. The dielectric strength, washing capability, density and viscosity of the machining fluid increase with an increase in emulsion concentration, pinch-effect and energy density of the discharge channel is enhanced, ejection effect of the eroded material increases; therefore, MRR rises. However, with a very high viscosity of the machining fluid, the eroded material is difficult to flush away, the stability of electrical discharges becomes bad; the MRR is low.

Fig. 1.19 shows the influence of emulsion concentration on SR. The SR increases with an increase in emulsion concentration, because the energy density of the discharge channel increases with an increase in emulsion concentration, the crater size generated by a single pulse becomes large; the SR rises with an increase in emulsion concentration.

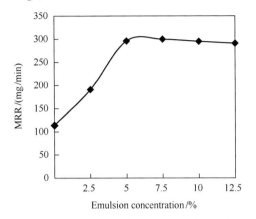

Fig. 1.18　Effect of emulsion concentration on MRR

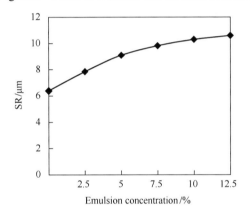

Fig. 1.19　Effect of emulsion concentration on SR

1.3.9　Effect of NaNO₃ concentration of the machining fluid on the process performance

The effect of $NaNO_3$ concentration on the MRR and the SR is illustrated in Fig. 1.20~Fig. 1.23, the machining fluid was 5% emulsion + water + $NaNO_3$.

Fig. 1.20 shows the relationship between the MRR and $NaNO_3$ concentration. The MRR initially rises with an increase in $NaNO_3$ concentration, and then decreases with an increase in $NaNO_3$ concentration. The reason for this is that dielectric strength of the machining fluid and breakdown delay time decreases with an increase in $NaNO_3$ concentration. It can be seen from Fig. 1.21 and Fig. 1.22 that the discharge time of a single pulse and the effective pulse frequency for 0.5% $NaNO_3$ concentration of the machining fluid are higher than that the machining fluid without $NaNO_3$; therefore, the MRR is high. As the $NaNO_3$ concentration is higher than a suitable value, electrolysis becomes strong, the gap voltage drop is great and electric discharges become weak, the MRR is low.

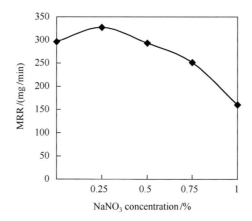

Fig. 1.20 Effect of $NaNO_3$ concentration on MRR

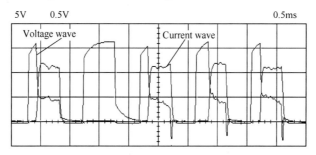

Fig. 1.21 Discharge waves with 5% emulsion + water

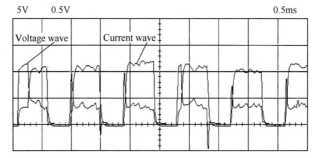

Fig. 1.22 Discharge waves with 5% emulsion + 0.5% $NaNO_3$ + water

As shown in Fig. 1.23, the SR initially increases with an increase in $NaNO_3$ concentration, and then decreases slowly with an increase in $NaNO_3$ concentration.

This is because that under a suitable NaNO$_3$ concentration, the discharge energy of a single pulse increases, the crater generated by a single pulse becomes large, the SR rises. As the NaNO$_3$ concentration is higher than a suitable value, the gap voltage drop is great and electric discharges become weak; the crater generated by a single pulse becomes small; the SR decreases.

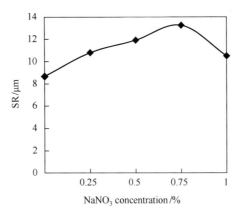

Fig. 1.23　Effect of NaNO$_3$ concentration on SR

1.3.10　Effect of Polyvinyl alcohol concentration on the process performance

The effect of polyvinyl alcohol (PVA) concentration on the MRR and the SR is illustrated respectively in Fig. 1.24 and Fig. 1.25, the machining fluid was 5% emulsion + water + PVA.

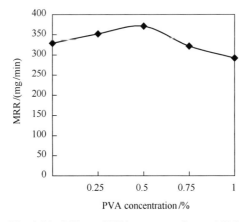

Fig. 1.24　Effect of PVA concentration on MRR

As shown in Fig. 1.24, the MRR initially increases with an increase in PVA concentration and then decreases with an increase in PVA concentration. The reason for this is that dielectric strength and viscosity of the machining fluid increase with an increase in PVA concentration, energy density of the discharge channel and discharge breakdown explosion force rise; the MRR is high. As PVA concentration is higher than a suitable value, the eroded material is difficult to flush away, the stability of electrical discharges becomes bad, the MRR is low.

Fig. 1.25 shows the influence of PVA concentration on SR. The SR increases with PVA concentration. This is because the energy density of the discharge channel and discharge breakdown explosion force increase with an increase in PVA concentration, the crater size generated by a single pulse becomes large. Fig. 1.26 and Fig. 1.27 show that the crater size generated by a single pulse with 0.5% PVA concentration of the machining fluid are higher than that for the machining fluid without PVA, therefore, the SR rises with an increase in PVA concentration.

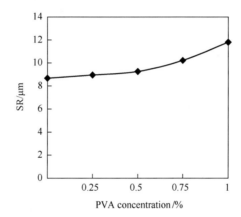

Fig. 1.25 Effect of PVA concentration on SR

Fig. 1.26 Surface photograph of the workpiece with 5% emulsion + water

Fig. 1.27 Surface photograph of the workpiece with 5% emulsion + 0.5%PVA + water

1.3.11 Effect of flow velocity of the machining fluid on the process performance

The effect of flow velocity on the MRR and the SR is illustrated respectively in Fig. 1.28 and Fig. 1.29; the machining fluid was 5% emulsion + water.

Fig. 1.28 shows the relationship between the MRR and flow velocity of the machining fluid. The MRR increases with flow velocity of the machining fluid. There are many reasons causing the phenomena. Machining fluids are flushed into the gap between the workpiece and the tool electrode at high speed, the eroded material are easy to be flush away, the stability of electrical discharges improves. The high flow velocity of the machining fluid make the assisting electrode becomes more close to the surface of the workpiece; electrical discharge energy acting on the surface of workpiece increases, the material removed by electrical discharge rises. Therefore, the MRR increases with flow velocity of the machining fluid.

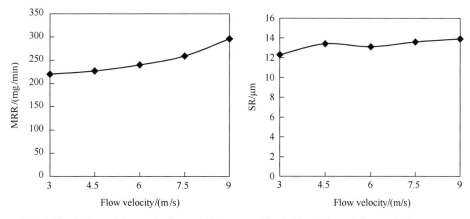

Fig. 1.28 Effect of flow velocity on MRR   Fig. 1.29 Effect of flow velocity on SR

As shown in Fig. 1.29, the SR increases slowly with an increase in flow velocity of the machining fluid. The reason for this is that the high flow velocity of the machining fluid make the assisting electrode becomes more close to the surface of the workpiece, electrical discharge energy acting on the surface of workpiece increases, the crater size generated by electric discharge becomes large, the SR increases.

1.4 Numerical simulation of single pulse discharge machining insulating Al$_2$O$_3$ ceramic

1.4.1 Model details

1. Physical model

Fig. 1.30 shows the physical model of machining insulating ceramics with single pulse discharge. During machining the copper electrode and the steel electrode are connected to the positive and negative poles of the power supply respectively. The power supply is able to generate a single pulse that the pulse duration can be adjusted continuously. The copper electrode is a thin sheet, it is placed on the insulating ceramic workpiece. The steel electrode is a needle-shaped electrode, it is fed towards the copper electrode along the surface of insulating ceramic workpiece, driven by a D. C. servo motor. As the distance between the steel needle electrode and the copper sheet electrode reaches the discharge gap, electrical discharges are produced. Because the copper sheet electrode is very thin and close to workpiece, most of electrical discharge energy acting on the assisting electrode and plasma energy can be directly acted on the workpiece surface. The instantaneous high temperature and pressure generated by electrical discharge cause the insulating ceramic to be removed.

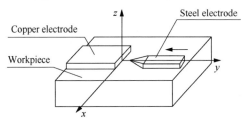

Fig. 1.30 Physical model of machining insulating ceramics with single pulse discharge

2. Thermal model

The three-dimensional unsteady-state heat conduction equation of the insulating ceramic, copper electrode and steel electrode is described as follows [24].

$$\rho(T)c(T)\frac{\partial T}{\partial t} = \lambda(T)\left[\frac{\partial}{\partial x}\left(\frac{\partial T}{\partial x}\right) + \frac{\partial}{\partial y}\left(\frac{\partial T}{\partial y}\right) + \frac{\partial}{\partial z}\left(\frac{\partial T}{\partial z}\right)\right] \quad (1.3)$$

Where, $\rho(T)$ is density, $c(T)$ is specific heat, $\lambda(T)$ is thermal conductivity, T is temperature.

The boundary conditions are

$$-\lambda\frac{\partial T}{\partial z} = \begin{cases} q & (|x| \leq R) \\ h(T-T_0) & (|x| > R) \end{cases} \quad (1.4)$$

Where, T_0 is ambient temperature, R is plasma radius, q is heat flux entering into the workpiece, h is heat transfer coefficient, x is distance from the heat source center.

The ambient temperature T_0 can be taken as room temperature.

$$T(x, y, z)_{t=0} = T_0 \quad (1.5)$$

Where, t is time.

3. Heat input

Insulating Al_2O_3 ceramic removal in EDM is through melting, vaporization, and spalling at the workpiece surface. This happens as a result of the heat flux that is applied on the workpiece by the electrical discharge channel between the copper electrode and the steel electrode. The high instantaneous temperature and pressure plasma on the workpiece surface are produced by the electrical discharge channel of flowing the insulating Al_2O_3 ceramic surface. Many researchers have considered uniformly distributed heat source within a spark [25,26]. This assumption is far from reality. DiBitonto et al. [27], and Bhattacharya et al. [28] had shown that Gaussian heat distribution is more realistic than disc heat source. Moreover, the assumption of Gaussian distribution is well-accepted for modeling the heat input in EDM. In this current work, the Gaussian heat input model is used to approximate the heat from the electrical discharge channel [29,30]. If it is assumed that total power of each pulse is to be used only by one spark, and the maximum intensity at the axis of a spark and its radius (R) are known, then the heat flux [$q(r)$] at radius r is given as follows:

$$q(r) = \frac{4.45 U_b I}{\pi R^2}\eta \exp\left\{-4.5\left(\frac{r}{R}\right)^2\right\} \quad (1.6)$$

Where, $q(r)$ is heat flux at radius r, U_b is breakdown voltage, I is discharge current, η is heat distribution coefficient, R is plasma radius, r is plasma radius.

A constant fraction of discharge energy is transferred to the ceramics, cathode electrode and anode electrode through the experiments. In the present work, energy transferred to ceramic is taken as 0.3, energy transferred to anode electrode is taken as 0.4, and energy transferred to cathode electrode is taken as 0.3.

The electrical discharge channel of flowing the insulating Al_2O_3 ceramic surface causes the ceramic to be removed. The crater on the workpiece surface is produced by electrical discharge channel. Width of the crater is related by the radius of electrical discharge channel. The discharge radius is related to the current intensity and pulse duration as in Erden's Equation [31,32]. According to the Erden's Equation and this experiment results, it is assumed that width of the crater produced by EDM is equal to the diameter of electrical discharge channel, the spark radius R as a function of pulse duration and current is written as

$$R = 0.0045 t_{on}^{0.2898} I^{0.71} \qquad (1.7)$$

Where, I is discharge current, t_{on} is pulse duration.

4. Phase change

When the temperature of the material exceeds the melting and evaporation point, latent heat due to melting and evaporation is needed. The latent heat required for phase change is accounted by modifying the expression for the specific heat. Considering enthalpy before and after phase change, the latent heat of melting and evaporation is incorporated into modified specific heat in the vicinity of melting temperature and boiling temperature [33].

$$c_m = c + \frac{L_m}{2\Delta T} \quad \text{for} \quad T_m - \Delta T \leqslant T \leqslant T_m + \Delta T \qquad (1.8)$$

$$c_v = c_m + \frac{L_v}{2\Delta T} \quad \text{for} \quad T_v - \Delta T \leqslant T \leqslant T_v + \Delta T \qquad (1.9)$$

Where, c_m is specific heat in liquid state, c_v is specific heat in vapour state, L_m is latent heat of melting, L_v is latent heat of evaporation, T_m is melting temperature, T_v is boiling temperature, c is specific heat, ΔT is change of temperature.

1.4.2 Finite element formulations

The Galerkin finite element formulation has been applied to obtain temperature distribution within the square domain due to electrical discharge. The following expressions are obtained:

$$\begin{aligned}
[C]^e &= \iiint_V \rho C \{N\}^e \{N\}^{eT} dxdydz \\
[K]^e &= \iiint_V K[B]^{eT}[B]^e dxdydz \\
\{f\}^e &= \iiint_V \{N\}^e \{N\}^{eT} \{q_w\}^e dxdydz
\end{aligned} \quad (1.10)$$

Where, $[C]^e$ is elemental capacitance matrix, $\{N\}^e$ is nodal shape function vector, $[K]^e$ is conductivity matrix, $[B]^e$ matrix relating temperature derivative with its nodal values, $\{f\}^e$ is boundary flux vector, $\{q_w\}^e$ is vector of nodal heat generation, $\{N\}^{eT}$ is transpose of nodal shape function vector.

The integrals in the Equation (1.10) are computed numerically using Gaussian quadrature with three points in each direction. When the elemental quantities of Equation (1.10) are assembled, the following differential equations are obtained:

$$[C(T)]\{\dot{T}\} + [K(T)]\{T\} = \{Q(T)\} \quad (1.11)$$

Where, $[C(T)]$ is global capacitance matrix, $\{T\}$ is global temperature vector, $\{\dot{T}\}$ is time derivative of $\{T\}$, $[K(T)]$ is global conductivity matrix, $\{Q(T)\}$ is global right side force vector.

Equation (1.11) represents a set of ordinary differential equations in the variable $\{T\}$ as a function of time t. These equations are converted to a set of algebraic equations with implicit and backward finite difference method given as follows.

$$\left[\frac{C(T)}{\Delta t} + [K(T)]\right]\{T\}_t = [Q(T)]_t + \frac{[C(T)]}{\Delta t}\{T\}_{t-\Delta t} \quad (1.12)$$

Where, $c(T)$ is specific heat, $[K(T)]$ is global conductivity matrix, $\{T\}$ is global temperature vector, $\{Q(T)\}$ is global right side force vector, t is time, $C(T)$ is specific heat capacity.

In each time interval Δt, the nodal temperatures at the new time level are calculated by solving the set of simultaneous algebraic Equation (1.12). The solution thus marches in time, in steps of Δt until the desired final time is reached.

1.4.3 Results and discussion

In the following simulations and experiments, the workpiece material is insulating Al_2O_3 ceramic, the negative electrode is a thin copper sheet, the positive electrode is a steel needle, the pulse time is 100μs, the peak current is 25A, and the machining fluid is water-based emulsion. The surface roughness is measured by a surface roughness tester. Single discharge crater parameters are measured by a universal tools microscope with computer data processing. Specific crater geometry analyses of the insulating Al_2O_3 ceramic is accomplished with the scanning electron microscopy (SEM).

1. Material properties

During machining temperature of the insulating Al_2O_3 ceramic surface where the discharge channels passes the workpiece goes from room temperature to the boiling point of the material and back to room temperature in a matter of a few microseconds. During this temperature changing, the insulating Al_2O_3 ceramic melts and vaporizes and some of the molten material re-solidifies. Thus, latent heat exchange is also an integral part of the process. It is imperative that temperature-dependent thermophysical property be used for the simulation process. Properties used in the current work are listed in Table 1.1~Table 1.3 [34,35].

Table 1.1 Properties of Al_2O_3 ceramic and copper

Material	Density ρ/(kg/m^3)	Melting point/℃	Boiling point/℃	Entropy (S)/[J/(mol·℃)]		
				Solid	Liquid	Gas
Al_2O_3	3600	2050	2980	291.91	388.98	393.44
Copper	8930	1083	2595	74.050	107.07	213.94
Steel	7860	1535	2750	92.61	125.55	237.06

Table 1.2 Specific heat of Al_2O_3 ceramic and copper with different temperature

T/℃	Specific heat/[J/(kg·℃)]		
	Al_2O_3	Copper	Steel
25	774	382	445
127	943	398	488
327	1093	418	674
527	1180	432	788
927	1257	458	802
1327	1307	490	810
1727	1350	490	822
2127	1420	490	822
2595	1420	382	821

Table 1.3 Thermal conductivity of Al$_2$O$_3$ ceramic and copper with different temperature

$T/°C$	Thermal conductivity (λ)/[W/(m · °C)]		
	Al$_2$O$_3$	Copper	Steel
100	31	482	48
200	26.38	413	46
446	12.59	393	41
646	8.898	379	35
846	7.079	366	25
1046	5.796	352	26
2000	4.946	245	29
2400	4.313	213	29
2868	3.752	182	29

2. Numerical simulation

Fig. 1.31 shows temperature fields generated by a single pulse discharge for insulating Al$_2$O$_3$ ceramic workpiece. The maximum temperature is located at the discharge channel center, it is 6445°C. The materials in discharge channel center area are mainly removed by evaporation. Temperature of the Al$_2$O$_3$ ceramic at molten region achieves the melting point. For the molten areas the phase changes from solid to liquid. Temperature of the workpiece at the heat-affected region does not reach the melting point of Al$_2$O$_3$ ceramic. Fig. 1.32 shows temperature fields due to single pulse discharge for section of insulating Al$_2$O$_3$ ceramic workpiece.

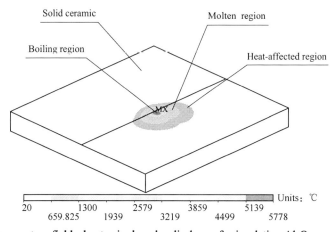

Fig. 1.31 Temperature fields due to single pulse discharge for insulating Al$_2$O$_3$ ceramic workpiece

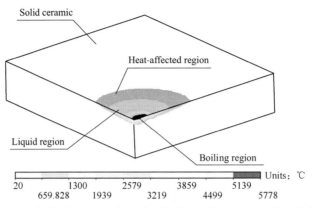

Fig. 1.32 Temperature fields due to single pulse discharge for section of insulating Al_2O_3 ceramic workpiece

Fig. 1.33 shows the variation of temperature along the distance in x-axis with the different depth from top surface (z-value). The temperature on the workpiece top surface with z=0 is highest. The more absolute z-value the temperature is lower. The temperature decreases with an increase in absolute x-value. As the x-value is equal to 0mm, the temperature is highest. On the workpiece top surface the ceramics at the area with $-0.118\text{mm} \leqslant x \leqslant 0.118\text{mm}$ are removed by boiling method, the ceramics at the region with $-0.186\text{mm} \leqslant x \leqslant -0.118\text{mm}$, $0.118\text{mm} \leqslant x \leqslant 0.186\text{mm}$ are removed by melting method. On the workpiece top surface, the material removal area with a single pulse discharge is $-0.186\text{mm} \leqslant x \leqslant 0.186\text{mm}$.

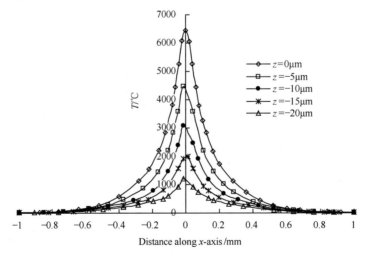

Fig. 1.33 Variation of temperature along distance in x-axis with different depth from top surface

As shown in Fig. 1.34, the temperature decreases with an increase in the absolute

z-value. As z-value is equal to 0μm, the temperature on the workpiece top surface is highest. The bigger absolute z-value, the temperature is lower. As the y-value is equal to 0.008mm, the temperature is highest. On the workpiece top surface, the ceramics at the zone with $-0.045\text{mm} \leqslant y \leqslant 0.125\text{mm}$ are removed by boiling method, the ceramics at the region with $-0.067\text{mm} \leqslant y \leqslant -0.045\text{mm}$, $0.125\text{mm} \leqslant y \leqslant 0.162\text{mm}$ are removed by melting method. On the workpiece top surface the material removal area with a single pulse discharge is $-0.067\text{mm} \leqslant y \leqslant 0.162\text{mm}$.

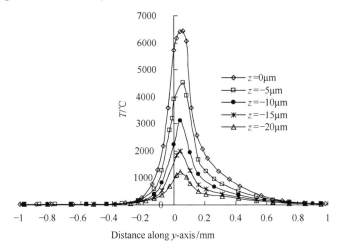

Fig. 1.34 Variation of temperature along distance in y-axis with different depth from top surface

3. Validation

As shown in Fig. 1.35, as x-value is equal to 0mm, the widths in y direction of experimental results and analytical results are maximums. The maximum width in y direction of experimental results is 0.318mm. The maximum width in y direction of analytical results is 0.280mm. As x is equal to ± 0.180mm, widths in y direction of numerical simulation results are equal to 0mm. As x is equal to ± 0.205mm, widths in y direction of experimental results are equal to 0mm. The simulation results are always less than experimental results. There are many reasons causing this phenomenon. The mathematical models of thermal eroding insulating Al_2O_3 ceramic with a single pulse discharge are only considered the material removal effects of melting and evaporation. The material removal of spalling effect is not considered in these models. However, there is material removal of spalling effect during EDM of insulating Al_2O_3 ceramic. Fig. 1.36 shows a typical area of material detachment on the insulating Al_2O_3 ceramic surface caused by the spalling mechanism, there are some potential flakes along the leading edges of the spalled area. These peripheral edges had an overhung appearance where horizontal cracking undermined the surface layer as it progressed into the material. Examination of the freshly exposed surface on the parent material revealed curved protruding edges indicative of propagating crack fronts running outwards from a central region within the spall area. The amount of energy required to remove material from the insulating Al_2O_3 ceramic surface by flake detachment is less than that required to remove material through melting and evaporation. Therefore, the simulation results are always less than experimental results.

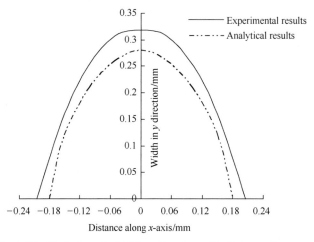

Fig. 1.35 Variation of width in y direction with distance along x-axis

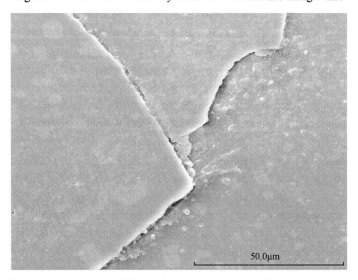

Fig. 1.36 SEM micrograph of Al_2O_3 ceramic surface spalled by single pulse discharge

Fig. 1.37 shows the width in x direction with distance along y-axis. As y-value is equal to 0.008mm, the widths in x direction of experimental results and analytical results are maximums. The maximum width in x direction of experimental results is 0.410mm. The maximum width in x direction of simulation results is 0.360mm. As y is equal to = −0.071mm or 0.087mm, widths in x direction of experimental results are equal to 0mm. As x is equal to −0.064mm or 0.08mm, widths in x direction of analytical results are equal to 0mm. The simulation results are always less than experimental results. The reason causing this is that analytical results are only considered that the ceramics are removed by melting and evaporation effects. The spalling effect is not considered in the analytical results. There are some ceramics are removed by spalling in actual machining process.

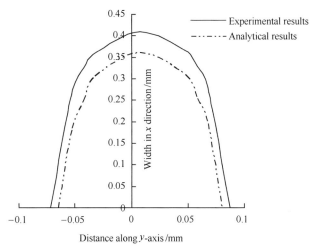

Fig. 1.37 Variation of length in x direction with distance along y-axis

1.5 Conclusions

Using a thin sheet copper close to the surface of insulating ceramics as the assisting electrode, insulating ceramics can be easily machined by ED milling. The process improves the stability of electrical discharges and shows high MRR. Using a water-based emulsion as the work fluid, harmful gas is not generated during DE milling, and its equipment is not corroded. During machining, positive polarity for the tool electrode should be used. Employing a conducting abrasive wheel instead of the steel tool electrode used in Ed milling, the process of finish compound grinding insulating ceramics can be easily done.

With a suitable emulsion concentration, the high machining rate and good surface quality of machining insulating Al_2O_3 ceramic by ED milling can be easily obtained. Using a suitable chemical additive and its dosage for water-based emulsion, dielectric strength, washing capability and viscosity of the machining fluid can be modified in order to increase the MRR of ED milling. Using high flow velocity of the machining fluid, the MRR of ED milling increases and its SR changes small.

A finite element based models for prediction of three-dimensional temperature fields on insulating ceramics generated by single pulse discharge are presented. Using these models the crater dimensions of insulating ceramics produced by single pulse discharge can be usefully predicted. The temperature at the discharge channel center is highest. The ceramics in discharge channel center area are mainly removed by evaporation. The temperatures in x, y, z directions decrease with an increase in absolute x-value, y-value, and z-value. The numerical simulation results are less than experimental results. The research results show that insulating Al_2O_3 ceramic are removed by single pulse discharge due to melting, evaporation, and spalling effects. The efficiency of spalling effect is higher than that of melting, and evaporation effects. The spalling effect is very difficult to be considered for the models of thermal eroding insulating Al_2O_3 ceramic with a single pulse discharge. This is because the material removal of spalling is a random process. The expression of the material removal by spalling with electrical discharge parameters is very difficult to be obtained.

References

[1] Anon. USA advanced ceramics demand. Materials Technology, 2005, 20(3): 168.
[2] Anon. Engineering ceramics market outlook. American Ceramic Society Bulletin, 2003, 82(11): 2.
[3] Yao S W, Wang H, Wang S L. The study and application of the special ceramics. Industrial Heating, 2006, 35(5): 1-4.
[4] Chisato T, Keisaku O, Tetsuya S. High quality machining of ceramics. Journal of Materials Processing Technology, 1993, 37(1-4): 639-654.
[5] Du J H, Liu Y H, Li X P, et al. The grinding technology of engineering ceramics. Materials for Mechanical Engineering, 2005, 29(3): 1-6.
[6] Mohri N, Fukuzawa Y T, Tani, et al. Some considerations to machining characteristics of insulating ceramics-towards practical use in industry. Annals of the CIRP, 2002, 51(1): 161-164.
[7] Li X P, Liu Y H, Ji R J, et al. Non-traditional machining technique for insulating engineering ceramics. Electromachining & Mould, 2006, (5): 6-9.
[8] Gopal A V, Rao P V. Modeling of grinding of silicon carbide with diamond wheels. Mineral Processing and Extractive Metallurgy Review, 2002, 23(1): 51-63.
[9] Agarwal S, Venkateswara R P. A new surface roughness prediction model for ceramic grinding. Proceedings of the Institution of Mechanical Engineers, Part B: Journal of Engineering

Manufacture, 2005, 219(11): 811-821.
[10] Liu W X. Research of relations among residual grinding stress, grinding condition and workpiece property on engineering ceramic ground surface. Journal of Hunan Institute of Science and Technology(Natural Science), 2006, 19(3): 64-67.
[11] Jia Z X, Wang Z Y, Guo Z J. Study on the surface quality of Si_3N_4 ceramics machined by wire EDM. Metallurgical Equipment, 2005, (4): 1-7.
[12] Kozak J, Rajurkar K P, Chandarana N. Machining of low electrical conductive materials by wire electrical discharge machining (WEDM). Journal of Materials Processing Technology, 2004, 149(1-3): 266-271.
[13] Lauwers B, Kruth J P, Brans K. Development of technology and strategies for the machining of ceramic components by sinking and milling EDM. CIRP Annals -Manufacturing Technology, 2007, 56(1): 225-228.
[14] Liu Y H, Guo Y F, Liu J C. Electric discharge milling of polycrystalline diamond. Proceedings of the Institution of Mechanical Engineers, Part B: Journal of Engineering Manufacture, 1997, 211(B8): 643-647.
[15] Guo Y H, Bai J C, Deng G Q, et al. High speed wire electrical discharge machining (HS-WEDM) phenomena of insulating Si_3N_4 ceramics with assisting electrode. Key Engineering Materials, 2007, 339: 281-285.
[16] Fukuzawa Y, Gotoh H, Mohri N, et al. Line swept surface generation on insulating ceramics by wire electrical discharge machining. Journal of the Australasian Ceramic Society, 2005, 4(1): 17-21.
[17] Liu Y H, Jia Z X, Liu J C. Study on hole machining of non-conducting ceramics by gas-filled electrodischarge and electrochemical compound machining. Journal of Materials Processing Technology, 1997, 69(1-3): 198-202.
[18] Bhattacharyya B, Doloi B N. Experimental investigations into electrochemical discharge machining (ECDM) of non-conductive ceramic materials. Journal of Materials Processing Technology, 1999, 95(1-3): 145-154.
[19] Guo Y H, Bai J C, Liu H S, et al. The Study of the process of machining insulating ceramics by electrical discharge grilling. Electromachining & Mould, 2006, (1): 54-64.
[20] Tani T, Fukuzawa Y, Furutani K, et al. Machining process of insulating ceramics by electrical discharge machining. Journal of the Japan Society for Precision Engineering, 1997, 63(9): 1310-1314.
[21] Fukuzawa Y, Tani T, Mohri N, et al. Micro hole electrical discharge machining on insulating ceramics with pipe electrode. International Journal of Electrical Machining, 2003, (8): 57-61.
[22] Nancy F P, Ahmed M G. Electrical discharge machining of advanced ceramics. Ceramic Bullentin, 1988, 67(6): 1048-1052.
[23] Liu J C, Zhao J Q. Non-conventional machining. Beijing: Machine Industry Press, 2005.
[24] Carslaw H S, Jaeger J C. Conduction of Heat in Solids, 2nd Edition. Oxford: Oxford University Press, 1959.
[25] Jilani S T, Pandey P C. Analysis and modeling of EDM parameters. Precision Engineering, 1982,

4(4): 215-221.

[26] Snoyes R, Van D F. Investigations of EDM operations by means of thermo mathematical models. Annals of CIRP, 1971, 20(1): 35.

[27] DiBitonto D D, Eubank P T, Patel M R, et al. Theoretical models of the electrical discharge machining process. I. A simple cathode erosion model. Journal of Applied Physics, 1989, 66(9): 4095-4103.

[28] Bhattacharya R, Jain V K, Ghoshdastidar P S. Numerical simulation of thermal erosion in EDM process. IE (I) Journal-PR, 1996, 77: 13-19.

[29] Yadav V, Jain V K, Dixit P M. Thermal stresses due to electrical discharge machining. International Journal of Machine Tools & Manufacture, 2002, 42: 877-888.

[30] Bhondwe K L, Yadava V, Kathiresan G. Finite element prediction of material removal rate due to electro-chemical spark machining. International Journal of Machine Tools & Manufacture, 2006, 46: 1699-1706.

[31] Ikai T, Hashigushi K. Heat input for crater formation in EDM // Proceedings of International Symposium for Electromachining-ISEM XI, EPFL, Lausanne, 1995: 163-170.

[32] Erden A. Effect of materials on the mechanism of electric discharge machining (EDM), Transactions ASME, Journal of Engineering Materials and Technology, 1983, 108: 247-251.

[33] Yadav V, Jain V K, Dixit P M. Temperature distribution during electro-discharge abrasive grinding. Machining Science and Technology, 2002, 6(1): 97-127.

[34] Yang S M. Basis of Heat Transfer. Beijing: High Education Press, 1991.

[35] Liu H L, Jin Z H, Hao Z Y, et al. Design, fabrication and properties of carbon/carbon/Al_2O_3 ceramic functionally gradient materials. Journal of Functional Materials and Devices, 2004, 10(1): 123-127.

Chapter 2 Single Discharge Machining of Insulating Ceramics Efficiently with High Energy Capacitor

Insulating ceramics are applied to modern manufacturing industries for their improved material properties. But they are the difficult-to-machine materials because of their high rigidity, high brittleness and non-electrical conductivity. A new method which employs a high energy capacitor for electric discharge machining of insulating ceramics efficiently is presented in this chapter, and the single discharge experiments have been carried out. The process uses the high voltage, large capacitor and high discharge energy, it is able to effectively machine insulating ceramics, and the single discharge crater volume of insulating ceramics can reach 17.63mm^3. The effects of polarity, peak voltage, capacitance, current-limiting resistance, tool electrode feed, tool electrode section area and assisting electrode thickness on the process performance such as the single discharge crater volume, the tool wear ratio and the assisting electrode wear ratio have been investigated. The microstructure of the discharge crater is examined with SEM. The results show that the discharge craters have sputtering appearance, the insulating ceramic materials are mostly removed by spalling, in the center region of the discharge some materials are removed by melting and vaporization, and the material removal is enhanced with the machining parameters increasing.

2.1 Introduction

Insulating ceramics have the advantages of good mechanical properties, light weight, high rigidity, chemical and thermal stabilities at elevated temperature and excellent wear resistance. Hence, they have been widely used in recent years in precision bearings, cutting tools, refractories, electronic components, automotive, aerospace and defense industries [1-3]. However, insulating ceramics are known as difficult-to-machine materials due to their beneficial properties such as high rigidity, high brittleness and non-electrical conductivity [4].

Grinding with diamond wheels is a major machining process for insulating ceramics, but it gives rise to difficulties, associated with the high cost of diamond wheels, large consumption of diamonds and laborious processing. Moreover, diamond grinding is characterized by significant mechanical and thermal impacts on the workpiece, resulting in the formation of splits and flaws on the cutting edge, and so resulting in a great number of defective and rejected components [5, 6]. Ultrasonic machining can be applied to machine various engineering ceramics due to its low cutting force and cutting heat, but the material removal rate is low and the tool wear is high [7, 8]. Laser beam machining is mostly used for cutting, drilling and carving the insulating ceramics, but it is costly, and micro-cracks will be created easily on the workpiece surface, which reduces the quality and life of the workpiece [9, 10]. EDM is a non-traditional machining process removing material by a succession of repeated electrical discharges between an electrode and a workpiece. Since the electrode does not contact with the workpiece during machining, EDM is an effective and economical technique for machining conducting ceramics with any rigidity. However, EDM cannot be directly used to machine insulating ceramics, since these materials are non-conducting. The development of the process for electro-machining insulating ceramics is an important research topic in the field of non-traditional machining. A lot of research achievements about the topic have been achieved [11-13].

Some researchers have used electrolyte as the machining fluid to achieve electrochemical discharge machining (ECDM) of insulating ceramics. This process can be applied to machine insulating materials, but it shows low efficiency, a lot of electric energy is wasted during machining, and the equipment can be eroded easily by electrolyte [14-16]. Fukuzawa et al. [17] and Muttamara et al. [18] have developed the techniques of the wire electrical discharge machining (WEDM) and EDM insulating ceramics using kerosene as machining fluid [17, 18]. In this method, a metal plate or metal mesh is arranged on the surface of the insulating ceramics as an assisting electrode. With the help of assisting electrode, insulating ceramics can be machined by sinking EDM or WEDM in kerosene. However, these processes of machining a large surface area on insulating ceramics show low efficiency. Akio has presented the MEEC process which integrates mechanical grinding, electrochemical machining and electrical discharge machining [19]. This process shows higher efficiency, better surface quality and lower cost compared to mechanical grinding, but a lot of electric energy is wasted during machining to produce the insulating gas used in electrical discharges, and the equipment can be easily eroded by electrolyte during machining. Liu et al. have developed a process of machining insulating ceramics using ED milling [20, 21]. ED milling uses a thin copper sheet fed to the steel wheel electrode along the surface of the workpiece as the assisting electrode and uses a water-based emulsion as the machining fluid. The process can machine insulating ceramics with high machining efficiency, low machining cost and non-pollution, and the material removal rate in machining insulating alumina can reach 133mm^3/min.

In order to further improve the machining efficiency, a new method which employs a high energy capacitor for electric discharge machining of insulating ceramics efficiently is presented in this chapter, and the single discharge experiments have been carried out. The results show that due to the high voltage, large capacitor and high discharge energy, insulating ceramics can be effectively machined by EDM, and the single discharge crater volume of insulating ceramics can reach 17.63mm^3. In this chapter, the effects of polarity, peak voltage, capacitance, current-limiting resistance, tool electrode feed, tool electrode section area and assisting electrode thickness on the process performance such as the single discharge crater volume (SDCV), tool wear ratio (TWR) and assisting electrode wear ratio (AEWR) have been investigated.

2.2 Experiments

2.2.1 Experimental principle

Fig. 2.1 shows the single discharge circuit of machining insulating ceramics with high energy capacitor. The circuit basically consists of a charging circuit and a discharging circuit. When the switch S_1 is turned on, the switch S_2 is turned off, and the capacitor C is charged. When the switch S_2 is turned on, the switch S_1 is turned off, the tool electrode is fed towards the assisting electrode put on insulating Al_2O_3 ceramic surface, driven by a D.C. servo motor. As the distance between the tool electrode and the assisting electrode reaches the discharge gap, electrical discharges are produced and the energy stored in the capacitor is directly discharged to the assisting electrode through the tool electrode. Because the sheet electrode is very thin and close to the workpiece, most of the electrical discharge energy acting on the assisting electrode and plasma energy can be directly acted on the workpiece surface. The instantaneous high temperature and plasma pressure cause melting, vapour, and spalling in the insulating ceramic. Fig. 2.2 is a single discharging photograph of machining insulating Al_2O_3 ceramic in air.

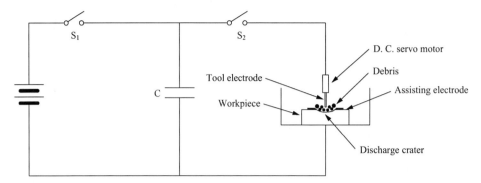

Fig. 2.1 Schematic illustration of single discharge machining insulating ceramics with high energy capacitor

Fig. 2.2 Single discharging photograph of machining insulating Al$_2$O$_3$ ceramic in air

2.2.2 Experimental procedure

In the following experiments, the workpiece material is insulating Al$_2$O$_3$ ceramic, the tool electrode material is red copper stick, the assisting electrode material is thin red copper sheet. The weighings of insulating Al$_2$O$_3$ ceramic, tool electrode and assisting electrode before and after machining are measured by an electronic balance (Sartorius BS224S). The TWR and AEWR are calculated according to Equation (2.1) and Equation (2.2), respectively. The microstructure of the discharge crater is observed by a scanning electron microscope (FEI QUANTA 200). The unspecified experimental parameters are summarized in Table 2.1 in the following experiments.

$$\text{TWR} = \frac{\rho_1 (m_2 - m_5)}{\rho_2 (m_1 - m_4)} \times 100\% \qquad (2.1)$$

$$\text{AEWR} = \frac{\rho_1 (m_3 - m_6)}{\rho_3 (m_1 - m_4)} \times 100\% \qquad (2.2)$$

Where, m_1, m_2, m_3 are the weighings of ceramic, tool electrode and assisting electrode before machining, respectively, m_4, m_5, m_6 are the weighings of ceramic, tool electrode and assisting electrode after machining, respectively, ρ_1, ρ_2, ρ_3 are the densities of ceramic, tool electrode and assisting electrode, respectively.

Table 2.1 Summary of experimental conditions

Polarity	Peak voltage/V	Capacitance /μF	Current-limiting resistance/Ω	Tool electrode section area/mm²	Assisting electrode thickness/mm	Tool electrode feed	Dielectric
Tool(−)	300	20000	0	2	0.05	vertical	air

2.3 Results and discussion

2.3.1 Effect of tool polarity on the process performance

Tool polarity is a primary factor that affects the process performance. The SDCV, TWR and AEWR with different tool polarities are given in Fig. 2.3(a)~Fig. 2.3(c), respectively.

The SDCV with different tool polarities is given in Fig. 2.3(a). Under the same conditions the SDCV in negative tool polarity is 1.1 to 1.7 times higher than that in positive tool polarity. Fig. 2.3(b) shows that under the same conditions the TWR in positive tool polarity is 3.3 to 4.3 times higher than that in negative tool polarity. According to the polarity effect, the workpiece is mostly affected by positive ions bombardment when the tool is connected to the positive pole, whereas the workpiece is mostly affected by electrons bombardment when the tool is connected to the negative pole [22]. During machining, as the distance between the tool electrode and the assisting electrode reaches the discharge gap, electrical discharge is produced, the energy stored in the capacitor is discharged to the discharge gap instantaneously, the discharge time is very short. Because the mass of the electrons is much smaller than that of positive ions, and they can be accelerated quickly during a short time, the bombardment effect by electrons is stronger than that by positive ions; the crater volume is high in negative tool polarity, TWR is high in positive tool polarity.

Fig. 2.3(c) shows the effect of tool polarity on the AEWR. Under the same conditions, the AEWR in positive tool polarity is 1.8 to 4.6 times higher than that in negative tool polarity. This phenomenon can be explained as follows. The crater generated by single discharge in negative tool polarity is deeper than that in positive tool polarity, the crater volume generated by single discharge in negative tool polarity is greater than that in positive tool polarity, under the same assisting electrode consumption, the ceramic material volume removed by single discharge in negative tool polarity is greater than that in positive tool polarity; therefore, the AEWR in positive tool polarity is high, negative tool polarity should be used.

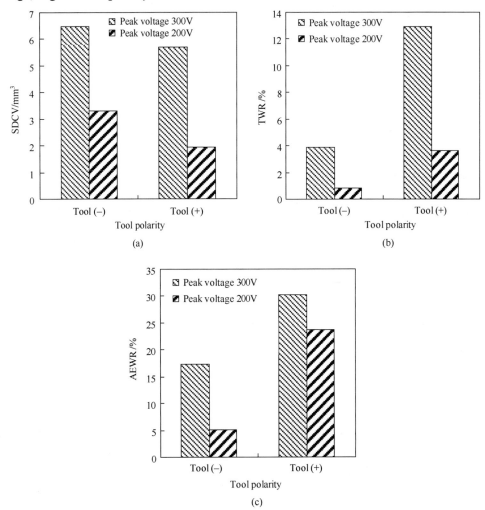

Fig. 2.3 Effect of tool polarity on the process performance. (a) Effect of tool polarity on SDCV, (b) Effect of tool polarity on TWR, (c) Effect of tool polarity on AEWR

2.3.2 Effect of peak voltage on the process performance

The effects of peak voltage on the SDCV, TWR and AEWR are illustrated in Fig. 2.4(a)~ Fig. 2.4(c), respectively. The SDCV, TWR and AEWR increase with an increase in peak voltage, respectively. The reason is that the energy stored in the capacitor increases with an increase in peak voltage, and the energy discharged to the electrode and the workpiece increases with an increase in peak voltage. It can also be seen from Fig. 2.4(b) that the TWR increases a little with the increase of peak voltage when the peak voltage is less than 300V, and it increases rapidly when the peak voltage increases from 300V to 350V. This phenomenon can be explained as follows. The single pulse energy and the tool wear increase with an increase in peak voltage, but the discharge crater also becomes larger, the interaction of the two functions makes the TWR change a little. But when the peak voltage increases from 300 V to 350 V, the discharge current and the thermal energy density increase rapidly, which enhances the discharge explosive force greatly and even causes the tool electrode material to be removed by flaking, therefore, the TWR increases rapidly.

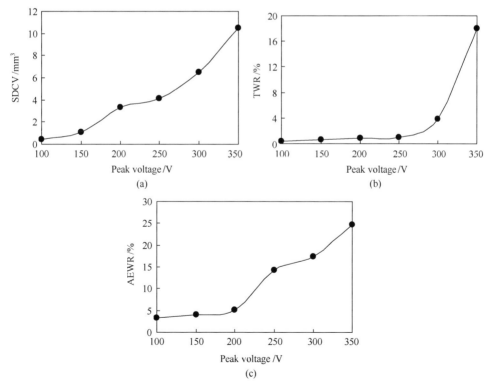

Fig. 2.4 Effect of peak voltage on the process performance. (a) Effect of peak voltage on SDCV, (b) Effect of peak voltage on TWR, (c) Effect of peak voltage on AEWR

2.3.3 Effect of capacitance on the process performance

Fig. 2.5 shows the effect of capacitance on the process performance. The SDCV, TWR and AEWR increase with an increase in capacitance, respectively. The reason is that the energy stored in the capacitor increases with an increase in capacitance, and the energy discharged to the electrode and the workpiece increases with an increase in capacitance. It can also be seen from Fig. 2.5(b) that the TWR increases a little with an increase in capacitance when the capacitance is smaller than 20000 μF, and it increases rapidly when the capacitance increases from 20000 μF to 25000 μF. The phenomenon can be explained as follows. The single pulse energy and the tool wear increase with an increase in capacitance, but the discharge crater also becomes larger, the interaction of the two functions makes the TWR change a little. But when the capacitance increases from 20000 μF to 25000 μF, the energy stored in the capacitor, the discharge current and the thermal energy density increase rapidly, which enhances the discharge explosive force greatly and even causes the tool electrode material to be removed by flaking, therefore, the TWR increases rapidly.

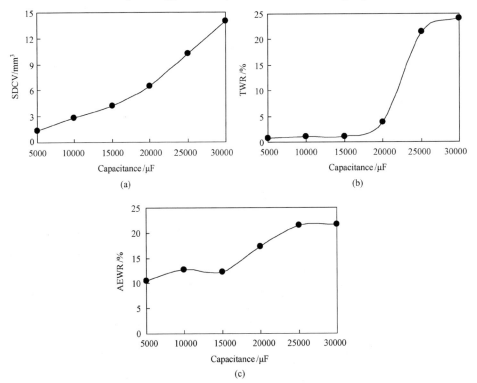

Fig. 2.5 Effect of capacitance on the process performance. (a) Effect of capacitance on SDCV, (b) Effect of capacitance on TWR, (c) Effect of capacitance on AEWR

2.3.4 Effect of current-limiting resistance on the process performance

The effect of current-limiting resistance on the process performance is shown in Fig. 2.6. TWR and AEWR decrease with an increase in current-limiting resistance, respectively. The reason is that the peak current and discharge energy decrease with an increase in current-limiting resistance. It can also be seen from Fig. 2.6(b) that the TWR is less than zero when the current-limiting resistance is greater than 4Ω. This is because the discharge energy and discharge explosive force decrease with the current-limiting resistance increasing, and some eroded material adheres to the tool electrode surface, which exceeds the tool electrode wear during machining.

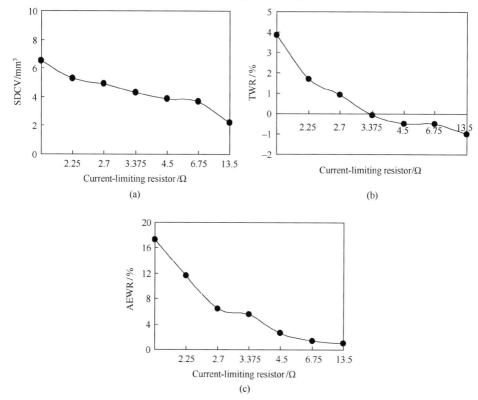

Fig. 2.6 Effect of current-limiting resistance on the process performance. (a) Effect of current-limiting resistance on SDCV, (b) Effect of current-limiting resistance on TWR, (c) Effect of current-limiting resistance on AEWR

2.3.5 Effect of tool electrode feed on the process performance

Figs. 2.7(a) and 2.7(b) show the illustrations of vertical tool electrode feed and level tool electrode feed, respectively, and Fig. 2.8 shows the effect of tool electrode feed on the process performance.

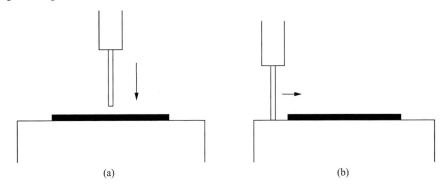

Fig. 2.7 Schematic illustration of different tool electrode feeds. (a) Vertical tool electrode, (b) Level tool electrode feed

The SDCV with different tool electrode feeds is given in Fig. 2.8(a). Under the same conditions the SDCV in level tool electrode feed is 1.3 to 1.5 times higher than that in vertical tool electrode feed. There are many reasons causing this phenomenon. When the tool electrode is fed vertically, the workpiece material is removed by the discharge energy transferred from the assisting electrode, whereas when the tool electrode is fed levelly, the discharge channel affects the workpiece surface directly, which enhances the workpiece material removal, so the SDCV is high in level tool electrode feed.

Fig. 2.8(b) shows the influence of tool electrode feed on the TWR. Under the same conditions the TWR in level tool electrode feed is 1.1 to 5.9 times higher than that in vertical tool electrode feed. This phenomenon can be explained as follows. The discharge space is smaller in vertical tool electrode feed than that in level tool electrode feed. The eroded material adheres to the tool electrode surface easily in vertical tool electrode feed, which can compensate the tool electrode wear, so the TWR is low in vertical tool electrode feed.

Fig. 2.8(c) shows that under the same conditions the AEWR in vertical tool electrode feed is 4.9 to 6.0 times higher than that in level tool electrode feed. The phenomenon can be explained as follows. The discharge energy diffuses around the discharge point on the assisting electrode surface in vertical tool electrode feed, whereas the discharge energy only acts in the thickness direction of the assisting electrode in level tool electrode feed. The heating assisting electrode area is large in vertical tool electrode feed, and more assisting electrode material can be removed, so the AEWR is high in vertical tool electrode feed.

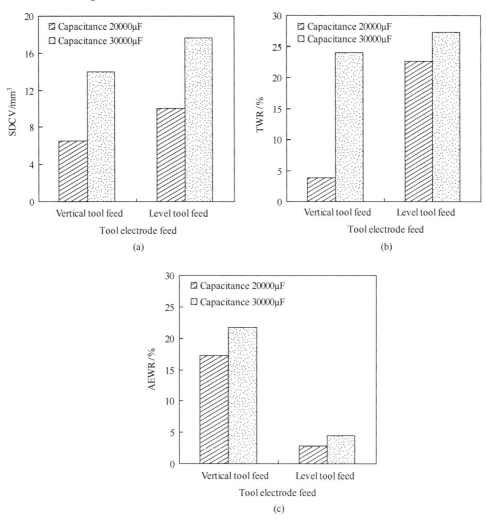

Fig. 2.8 Effect of tool electrode feed on the process performance. (a) Effect of tool electrode feed on SDCV, (b) Effect of tool electrode feed on TWR, (c) Effect of tool electrode feed on AEWR

2.3.6 Effect of tool electrode section area on the process performance

Fig. 2.9 shows the effect of tool electrode section area on the process performance. As shown in this figure, the SDCV, TWR and AEWR decrease with an increase in tool electrode section area, respectively. This is because that the energy density of discharge channel decreases with an increase in tool electrode section area, which reduces the ceramic and electrode material removal. It can also be seen from Fig. 2.9(b) that the TWR is less than zero when the tool electrode section area is greater than 4mm^2. This is because the discharge energy and discharge explosive force decrease with the tool electrode section area increasing, and some eroded material adheres to the tool electrode surface, which exceeds the tool electrode wear during machining.

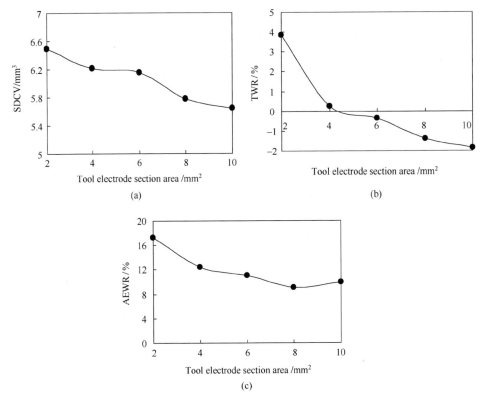

Fig. 2.9 Effect of tool electrode section area on the process performance. (a) Effect of tool electrode section area on SDCV, (b) Effect of tool electrode section area on TWR, (c) Effect of tool electrode section area on AEWR

2.3.7 Effect of assisting electrode thickness on the process performance

The effect of assisting electrode thickness on the process performance is illustrated in Fig. 2.10. The SDCV, TWR and AEWR decrease with an increase in the assisting electrode thickness, respectively. The phenomena can be explained as follows. The energy that is needed to erode the copper sheet increases with the assisting electrode thickness increasing, the energy that transfers to the workpiece surface decreases with an increase in the assisting electrode thickness, so the SDCV decreases. In addition, it can be observed from the experiment that the discharge spark and discharge sound decrease with an increase in assisting electrode thickness during machining, which means that the discharge explosive force decreases, the eroded electrode material can adhere to the electrode surface easily; therefore, the TWR and AEWR decrease.

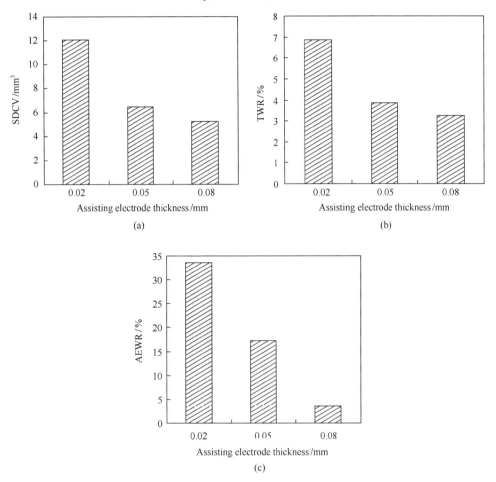

Fig. 2.10 Effect of assisting electrode thickness on the process performance. (a) Effect of assisting electrode thickness on SDCV, (b) Effect of assisting electrode thickness on TWR, (c) Effect of assisting electrode thickness on AEWR

2.4 Microstructure character of single discharge crater on insulating ceramic surface

Fig. 2.11 depicts the top surface morphology of the single pulse discharge craters of insulating ceramics with different single discharge energies. The craters illustrated in

Fig. 2.11(a) and Fig. 2.11(b) have sputtering appearance, and the sputtering effect intensifies with the single discharge energy increasing. There are many reasons causing the phenomena. As the pulse energy is discharged in a rather short time, a discharging micro-zone with high pressure and high temperature is produced, and the workpiece and electrode material in the discharging micro-zone will melt. The melting material gushes and splashes with the expansion of the discharge channel. After the pulsed discharge, the temperature of the original discharging micro-zone sharply decreases, resulting in rapid solidification of the gushed and splashed melting material. Thus, the single pulse sputtering appearance is produced. The melting material and discharge explosive force increase with an increase in single discharge energy, therefore, the sputtering effect intensifies.

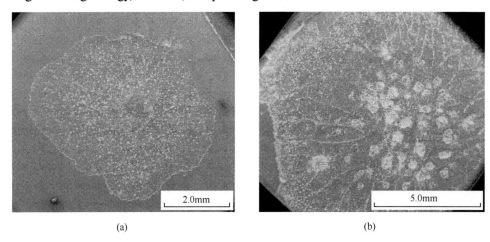

Fig. 2.11 SEM micrograph of discharge craters with different single discharge energies. (a) Single discharge energy of 225 J, (b) Single discharge energy of 1350 J

The microstructure of discharge crater bottom with different peak voltages is shown in Fig. 2.12. There are many small circular craters at the discharge crater bottom, and the number of small circular craters increases with the peak voltage increasing. The phenomena can be explained as follows. The high peak voltage and capacitance are used during machining, and the single discharge energy is so high that the ceramic material can melt and even vaporize rapidly. Moreover, the discharge channel deflects and fluctuates due to the plasma oscillation, and many small circular craters are produced at the discharge crater bottom. In addition, the discharge energy density increases with an increase in peak voltage, and the discharge channel deflects and fluctuates violently, so the number of small circular craters increases. It can also be seen from Fig. 2.12 that there are some angular grains left in the small circular craters, and the number of grains decreases with an increase in peak voltage. This is because that only part of melting material can be ejected, and the other cools and forms grain in the small circular crater. The discharge energy and discharge explosive force increase with the peak voltage increasing, and more melting material can be ejected, so the number of grains decreases with an increase in peak voltage.

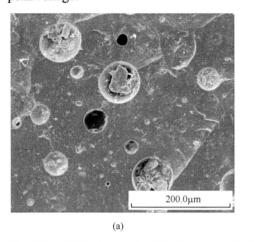
(a)

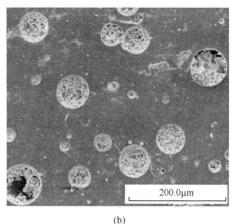
(b)

Fig. 2.12 SEM micrograph of discharge crater bottom with different peak voltages. (a) Peak voltage of 100V, (b) Peak voltage of 300V

Fig. 2.13 shows the microstructure of the discharge crater edge with different capacitances. As can be observed in this figure, the material is removed by spalling on the discharge crater edge, and the spalling effect intensifies with the capacitance increasing. There are many reasons causing the phenomena. The energy stored in the capacitor is high, and it is discharged to the discharge gap instantaneously during machining. With extreme hardness and brittleness, the insulating Al_2O_3 ceramic tends to promote the formation of steep temperature gradient and thermal stress at the workpiece discharge point. The exceedingly high thermal stress can exceed the maximum tensile strength of the ceramic easily, so the workpiece material is removed by spalling. The discharge energy and thermal stress increase with an increase in capacitance, therefore, the spalling effect intensifies.

The single discharge machining of insulating ceramics with high energy capacitor can be expressed as follows according to the above analysis. The discharge channel with high temperature is produced instantaneously after the dielectric breakdown, a lot of energy is transferred to the workpiece by the assisting electrode, and the high temperature gradient and thermal stress are induced on the workpiece surface, so the insulating ceramic materials are mostly removed by spalling, in the center region of the discharge some materials are removed by melting and vaporization. Because the melting material gushes and splashes with the expansion of the discharge channel, the discharge craters have sputtering appearance. In addition, there are many small circular craters left at the discharge crater bottom due to the deflection and fluctuation of the discharge channel.

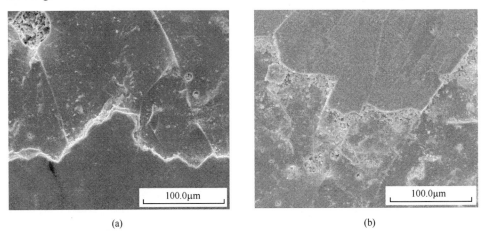

(a)   (b)

Fig. 2.13 SEM micrograph of discharge crater edge with different capacitances. (a) Capacitance of 5000 μF, (b) Capacitance of 30000 μF

2.5 Conclusions

A new method which employs a high energy capacitor for electric discharge machining of insulating ceramics efficiently is presented in this chapter, and the single discharge experiments have been done. The results show that due to the high voltage, large capacitor and high discharge energy, insulating ceramics can be effectively machined by EDM, and the single discharge crater volume of insulating ceramics can reach 17.63mm^3.

The SDCV, TWR and AEWR increase with the increase of peak voltage, capacitance, respectively, whereas they decrease with the increase of current-limiting resistance, tool electrode section area, and assisting electrode thickness, respectively. Higher SDCV, higher TWR and lower AEWR can be acquired with level tool electrode feed. Higher SDCV, lower TWR and lower AEWR can be acquired in negative tool polarity, so negative tool polarity is suitable for machining the insulating ceramics.

In the developed process the discharge craters on insulating ceramics have sputtering appearance, and there are many small circular craters left at the discharge crater bottom. The insulating ceramic materials are mostly removed by spalling, in the center region of the discharge some materials are removed by melting and vaporization, and the material removal is enhanced with the machining parameter increasing.

References

[1] Lorenzo M C, Ajayi O O, Singh D, et al. Friction and wear behavior of zirconia ceramic materials. Ceramic Engineering and Science Proceedings, 2009, 29(4): 75-84.
[2] Bandyopadhyay W S, Biswas S K, Maiti H S. Nitride & oxy-nitride ceramics for high temperature and engineering applications. Key Engineering Materials, 2009, 395: 193-208.
[3] Luo H H, Zhang F C, Wang T S. Prediction of fracture characteristic of particle-reinforced alumina-based composites. Science in China, Series E: Technological Sciences, 2009, 52(4): 864-870.
[4] Tian X L, Yang J F, Liu C, et al. Research progress of advanced machining technologies for engineering ceramics. Advanced Materials Research, 2009, 69/70: 359-363.
[5] Guo L, Xie G Z, Li B. Grinding temperature in high speed deep grinding of engineering ceramics. International Journal of Abrasive Technology, 2009, 2(3): 245-258.
[6] Arai S, Wilson S A, Corbett J, et al. Ultra-precision grinding of PZT ceramics-surface integrity control and tooling design. International Journal of Machine Tools and Manufacture, 2009, 49(12/13): 998-1007.
[7] Mehta R C S, Jadoun R S, Kumar P, et al. Application of Taguchi method in the optimization of process parameters for conicity of holes in ultrasonic drilling of engineering ceramics. Ceramic Engineering and Science Proceedings, 2008, 28(7): 167-178.
[8] Churi N J, Pei Z J, Shorter D C, et al. Rotary ultrasonic machining of dental ceramics. International Journal of Machining and Machinability of Materials, 2009, 6(3/4): 270-284.
[9] Samant A N, Dahotre N B. Laser machining of structural ceramics: A review. Journal of the European Ceramic Society, 2009, 29(6): 969-993.
[10] Kacar E, Mutlu M, Akman E, et al. Characterization of the drilling alumina ceramic using Nd: YAG pulsed laser. Journal of Materials Processing Technology, 2009, 209(4): 2008-2014.
[11] Guo Y F, Deng G Q, Bai J C, et al. Electrical Discharge Machining (EDM) phenomena of insulating ZrO_2 ceramics with assisting electrode. Key Engineering Materials, 2008, 375/376: 313-317.
[12] Liu Y H, Ji R J, Li X P, et al. Effect of machining fluid on the process performance of electric discharge milling of insulating Al_2O_3 ceramic. International Journal of Machine Tools and Manufacture, 2008, 48(9): 1030-1035.
[13] Yin S H, Ohmori H, Dai Y T, et al. ELID grinding characteristics of glass-ceramic materials. International Journal of Machine Tools and Manufacture, 2009, 49(3/4): 333-338.
[14] Chak S K, Venkateswara R P. Trepanning of Al_2O_3 by electro-chemical discharge machining (ECDM) process using abrasive electrode with pulsed DC supply. International Journal of

Machine Tools and Manufacture, 2007, 47(14): 2061-2070.
[15] Liu Y H, Jia Z X, Liu J C. Study on hole machining of nonconducting ceramics by gas-filled electrodischarge and electrochemical compound machining. Journal of Materials Processing Technology, 1997, 69(1-3): 198-202.
[16] Chak S K, Venkateswara R P. The drilling of Al_2O_3 using a pulsed DC supply with a rotary abrasive electrode by the electrochemical discharge process. International Journal of Advanced Manufacturing Technology, 2008, 39(7/8): 633-641.
[17] Fukuzawa Y, Mohri N, Tani T. Machining characteristics of insulating ceramics by electrical discharge machine. Industrial Ceramics, 2001, 21(3): 187-189.
[18] Muttamara A, Fukuzawa Y, Mohri N, et al. Probability of precision micro-machining of insulating Si_3N_4 ceramics by EDM. Journal of Materials Processing Technology, 2003, 140(1-3): 243-247.
[19] Akio K. MEEC machining of new materials and difficult-to-machine materials. Machine Tools, 1986, 24(12): 82-88.
[20] Liu Y H, Ji R J, Li X P, et al. Electric discharge milling of insulating ceramics. Proceedings of the Institution of Mechanical Engineers, Part B: Journal of Engineering Manufacture, 2008, 222(2): 361-366.
[21] Liu Y H, Li X P, Ji R J, et al. Effect of technological parameter on the process performance for electric discharge milling of insulating Al_2O_3 ceramic. Journal of Materials Processing Technology, 2008, 208(1-3): 245-250.
[22] Cao F G. Electrical discharge machining. Beijing: Chemical Industry Press, 2005: 16-18.

Chapter 3 Electrical Discharge Mechanical Grinding of Insulating Engineering Ceramics

A new method known as electrical discharge and grinding with synchronous servo double electrodes (EDGSSDE) is presented in this chapter for insulating engineering ceramics which integrates electrical discharge and mechanical grinding with a metal bonded diamond conductive wheel, the mechanism of this method is also analyzed. The results show that electrical discharge and electrolysis occur simultaneously during the process, which makes the diamond grains always in knife-edge, and achieves in-process dressing of grinding wheel. Due to the effects of high temperature and high heat generated by electrical discharge, a metamorphosed layer is formed on the surface of ceramics, which debases the rigidity of workpiece. The conductive grinding wheel grinds the metamorphosed layer and achieves paucity of ceramics removal by minute fritter. This method can dress grinding wheel in-process, avoid micro-crack that produced in traditional engineering grinding, reduce the grinding force, improve the service life of grinding wheel, advance the surface quality and machine insulating engineering ceramics in ultra-precision.

The non-uniform high temperature distribution of engineering ceramics generated during electrical discharge grinding results in residual stresses. These residual stresses can decrease the engineering ceramics in fatigue life and strength, even lead to micro-cracks. A finite element-based analysis of residual stresses in engineering ceramics electrical discharge grinding is presented. The temperature distribution of the ceramics during the machining has to be analyzed first by finite element method (FEM) for non-linear transient temperature field. Two different heat sources have to be developed due to the dissimilarity between the grinding heat source and the EDM heat source. The temperature distribution within ceramics due to grinding is determined by moving heat source. Gaussian heat distribution has been considered in the calculation of temperature distribution due to EDM. Temperature distribution in the ceramics due to electrical discharge grinding is obtained by using superposition. Then the residual stresses in the ceramics are determined by FEM for the thermal elastic-plasticity using the temperature distribution of electrical discharge grinding. The distribution of the temperature fields and residual stresses is analyzed. The analysis of residual stresses can predict the micro-cracks.

The effects of electrical parameters, such as pulse width, pulse interval, peak voltage, and peak current, on material removal rate and surface roughness of the insulating Al_2O_3 ceramics workpiece are discussed in detail. The results of the experiments reveal that higher MRR with acceptable surface quality can be obtained by this new electrical discharge grinding method with appropriate selection of parameters.

3.1 Introduction

In recent years, insulating engineering ceramics are widely used in precision bearings, cutting tools, refractories, electronic components, and in the automotive, aerospace, and defense industries [1]. The advantages of insulating engineering ceramics are high intensity, high rigidity, light weight, chemical stability, and superior wear-resistance. However, the cost of machining them is very high due to low removal rates, high super-abrasive wheel wear rates, expensive diamond grinding wheel and high-stiffness grinder [2]. Some studies suggest that the cost of grinding may account for up to 75% of the component costs for ceramics compared to 5%~15% for metallic components [3]. In addition, micro-cracks will be created easily on workpiece surface in grinding insulating engineering ceramics due to strong mechanical removal force between grinding wheel and workpiece, which reduces the quality and life of workpiece. Manufacturing engineers have tried to solve the problems in non-traditional machining ways such as ultrasonic machining, laser beam machining, assistant electrode EDM, compound electrochemical machining ECM/EDM, mechanical grinding, electrolysis electrical discharge compound machining MEEC, electrolytic in-process dressing (ELID) and so on [4-9]. These researches promote the developments of insulating engineering ceramics machining, but also bring the disadvantages of high cost, low efficiency and poor quality.

One of the present authors has pioneered a novel grinding technique known as EDGSSDE for insulating engineering ceramics, which can integrate the advantages of electrical discharge and mechanical grinding, achieve in-process dressing of grinding wheel and make the diamond grains always in knife-edge. By adjusting the parameters of discharge and grinding, rough, semi-finish and finish machining for insulating engineering ceramics can be achieved in the same machine tool, which can reduce the cost, improve the removal rate, advance the surface quality and machine insulating engineering ceramics in ultra-precision.

Surface integrity plays an important role in engineering ceramics. Surface cracks are aroused mainly by ceramics residual stresses from the EDG. Residual stresses can decrease the engineering ceramics in fatigue life and strength, even lead to micro-cracks. Theoretical study of residual stresses is necessary because purely experimental investigation is too expensive and sometimes difficult to perform. Available literature reveals that most of the residual stresses models are developed either on grinding or on EDM, residual stresses models of ceramics on EDG are scarce. Hamdi et al. [10] and Tian et al. [11] applied FEM to analysis the residual stresses due to grinding. The EDM residual stresses is analyzed by Das et al. [12] and Rebelo et al. [13]. Yadava et al. [14] develope a finite element method based mathematical model to simulate the compound machining process of grinding and EDM for thermal stresses in the workpiece. A review of the literature indicates that there have been very few theoretical approaches for determination of residual stresses in ceramics during EDG.

In this chapter, a model is presented based on the finite element method in engineering ceramics for the prediction of residual stresses occurring during EDG. In order to determine the residual stresses in the engineering ceramics during EDG, the temperature distribution has to be analyzed first. The temperature distribution within ceramics due to grinding is determined by moving heat source. Gaussian heat distribution has been considered in the calculation of temperature distribution due to EDM. Temperature distribution of engineering ceramics due to EDG is obtained by using superposing the two temperature distributions.

The authors have developed a new electrical discharge grinding method. It employs a conductive metal grinding wheel which rotates fast on the surface of workpiece, and a sheet electrode which is automatically fed to the place that the wheel and workpiece meet. The pulse power supply is applied to the sheet and wheel, and discharges between these two electrodes will happen when the voltage rises to some degree. Then the surface of workpiece can be machined continuously by the energy released from these discharges.

As a new processing method, electrical discharge grinding of insulating engineering ceramics has many differences with traditional EDM of metal workpieces. To further study the processing methods, and obtain optimized electrical parameters, the authors have researched effects of electrical parameters on MRR and SR through experiment. And the results are discussed in theory.

3.2 Principle of EDGSSDE

The principle of EDGSSDE is shown in Fig. 3.1, Discharge occurs between metal bond and assistant electrode when assistant electrode feeds on the workpiece surface in the process, then spark energy on the workpiece is utilized for machining ceramics. Due to the effects of re-solidified zone and high gradient temperature generated by electrical discharge, a metamorphosed layer is formed on the surface of ceramics, which has loose organization and low rigidity. The diamond grains grind the metamorphosed layer, achieve paucity of ceramics removal by minute fritter and improve the surface quality. In-process dressing of grinding wheel is achieved when machining ceramics, which makes the diamond grains always in knife-edge, and an oxide layer is formed on grinding wheel surface during the process, which can prevent excessive wear of grinding wheel, and advance the surface quality.

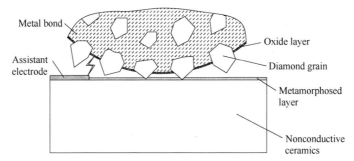

Fig. 3.1   Schematic of EDGSSDE process

3.3 Formation and function of oxide layer on grinding wheel

Cast iron bonded diamond conductive wheel is used in this technique. Fe_2O_3 and Fe_3O_4 are formed due to the effects of high heat generated by electrical discharge in the following chemical reactions:

$$4Fe + 3O_2 \longrightarrow 2Fe_2O_3$$

$$3Fe + 2O_2 \longrightarrow Fe_3O_4$$

Working fluid has determinate electrical conductivity, hence the bond will be removed by electrolysis, which is mostly ionized into Fe^{2+}. The Fe^{2+} will form $Fe(OH)_2$ or $Fe(OH)_3$ according to the following chemical reactions:

$$Fe - 2e \longrightarrow Fe^{2+}$$
$$Fe^{2+} - e \longrightarrow Fe^{3+}$$
$$H_2O \longrightarrow H^+ + OH^-$$
$$Fe^{2+} + 2OH^- \longrightarrow Fe(OH)_2 \downarrow$$
$$Fe^{3+} + 3OH^- \longrightarrow Fe(OH)_3 \downarrow$$

$Fe(OH)_2$ is oxidated into $Fe(OH)_3$ easily, and the hydroxides further change into oxides such as Fe_2O_3 in the following reactions:

$$4Fe(OH)_2 + O_2 + 2H_2O \longrightarrow 4Fe(OH)_3$$
$$2Fe(OH)_3 \longrightarrow Fe_2O_3 + 3H_2O$$

Thus the oxide layer on the wheel surface is composed of Fe_2O_3 and Fe_3O_4, and grows thicker gradually due to high temperature and electrolysis, but grows thinner due to scratching on grinding wheel surface by workpiece. The two functions will achieve homeostasis. The insulating oxide layer wears down in grinding workpiece, and will be recovered due to high temperature and electrolysis. Therefore, abrasive grains become protrudent, which achieves in-process dressing of grinding wheel, and machines insulating engineering ceramics in ultra-precision steadily.

The oxide layer on grinding wheel plays an important role in achieving ultra-precision grinding. First, it makes workpiece contactless with metal bond immediately, which can avoid the debasement of surface quality. Second, it can absorb and damp vibration partly because of the lower stiffness of oxide layer than that of workpiece, decrease vibratory shock and reduce friction. Third, it can also accomodate some abrasive grains generated from grinding wheel, which makes the wheel a grinding disc polishing the workpiece, and the abrasive grains are fixed in the oxide layer, which increases grinding function and advances grinding efficiency. When the abrasive grains on the wheel surface fall off due to extrusion and friction, the oxide layer wears down in time, which can avoid pluging grinding wheel, reduce the finishing time of grinding wheel, and improve grinding efficiency and stability.

3.4 Formation and function of metamorphosed layer on workpiece surface

The surface material of workpiece melts or boils away due to high heat generated by electrical discharge, and the left melting material forms a paper-thin metamorphosed layer due to the effects of finishment of discharge and cooling of working fluid. The rigidity of metamorphosed layer is lower than that of workpiece, which is in favor of the transition in material removal process from brittle to ductile, and achieves ductile grinding, and smooth surface.

The definition of ductile regime grinding according to Bifano is that cracked area of brittle material is lower than 10 percent of all machined area [15]. A mass of experiments and theoretical analysis show that transition in the removal process from brittle to ductile can achieve with proper machining condition, which is the maximal cutting depth of single grain less than critical penetration depth of brittle material.

The critical penetration depth a_c for fracture initiation is described as follows [16]:

$$a_c = 0.15 \left(\frac{E}{H}\right)\left(\frac{K_{Ic}}{H}\right)^2 \tag{3.1}$$

Where, E is the elastic modulus, H is the hardness, K_{Ic} is the fracture toughness, and b is a constant which depends on machining condition.

The contact state of grinding wheel and workpiece is analyzed and the maximal cutting depth a_{gmax} of single grain can be described as follows [17]:

$$a_{gmax} = \left(\frac{4v_w}{v_s N_d C}\sqrt{\frac{a_p}{d_c}}\right)^{\frac{1}{2}} \tag{3.2}$$

Where, v_w is the feedrate of workpiece, a_p is the grinding depth, v_s is the velocity of grinding wheel, N_d is the number of grinding grain in effect, d_c is the equivalent diameter, and C is a constant which depends on machining condition.

A metamorphosed layer is formed on the workpiece surface, which debases the rigidity H of workpiece, largens the critical penetration depth a_c, and the condition $a_{gmax} < a_c$ can easily be achieved. Furthermore, the grinding wheel cannot be plugged due to in-process dressing, which can make the diamond grains always in knife-edge, achieve paucity of ceramics removal by minute fritter, and machine insulating engineering ceramics in ultra-precision in finish machining.

3.5 Analysis of residual stresses in EDGSSDE of engineering ceramics

As discussed earlier, the goal of this chapter is to develop a model to predict residual stresses of ceramics occurring during EDGSSDE. Because the nature of EDGSSDE is random and complex, the following assumptions are made to make the analysis of residual stresses tractable.

3.5.1 Assumptions

The assumptions on analyis of residul stresses are as follows:
    (1) The engineering ceramics is homogeneous and isotropic.
    (2) The properties of engineering ceramics are temperature independent.
    (3) The heat transfer to the engineering ceramics is by conduction.
    (4) Inertia and body force effects are negligible during stress development.

3.5.2 Temperature models

Accurate prediction of residual stresses occurring during EDGSSDE requires determination of the temperature distribution in the ceramics at first. Due to the dissimilarity between the grinding heat source and the EDM heat source, two different heat sources have to been developed. Temperature distribution in ceramics due to EDGSSDE is obtained by superposition.

1. Temperature models of grinding

The temperature produced by a moving heat source in ceramics is analyzed. Two-dimensional thermal model of the grinding process by a rectangular distributed heat moving across the surface is made as shown in Fig. 3.2.

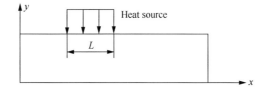

Fig. 3.2 Two-dimensional thermal model of the grinding

    Temperature field of ceramics due to grinding is governed by the following transient heat conduction equation:

$$\rho C_p \frac{\partial T}{\partial t} = \frac{\partial}{\partial x}\left(k_x \frac{\partial T}{\partial x}\right) + \frac{\partial}{\partial y}\left(k_y \frac{\partial T}{\partial y}\right) \tag{3.3}$$

Where, T is temperature, t is time, ρ is density, k_x, k_y are thermal conductivity in the x and y directions respectively, C_p is specific heat capacity of ceramics, x, y are coordinate axes.

The boundary conditions on the surface are

$$k_x \frac{\partial T}{\partial x} n_x + k_y \frac{\partial T}{\partial y} n_y + h(T - T_\infty) + q = 0 \tag{3.4}$$

Where, q is heat flux on the surface, h is the convection coefficient, k_x is the temperature coefficient in x direction, k_y is the temperature coefficient in y direction, n_x is the unit vector in x direction, n_y is the unit vector in y direction, T_∞ is the temperature at infinity.

To model Equation (3.3), the finite element method is used where the temperature T is assumed as the nodal degree of freedom. The η-family time integration scheme [18] is used to obtain the numerical solution for Equation (3.3). The finite element formulation of the heat-conduction equation for a typical element at time $(t+\eta t)$ can be written as

$$[C]^{t+\eta t}\left\{\frac{\partial T}{\partial t}\right\}^{t+\eta t} + [K]^{t+\eta t}\{T\}^{t+\eta t} + [H]^{t+\eta t}\{T\}^{t+\eta t} = -\{q\}^{t+\eta t} + [H]^{t+\eta t}\{T_\infty\} \tag{3.5}$$

Where, $[C]$ is the heat capacity matrix, $[K]$ is the conductivity matrix, $\{T\}$ is the vector of temperature, $\{T_\infty\}$ is the vector of the temperature at infinity, $[H]$ is the convection matrix, $\{q\}$ is the vector of heat flow inputs. The variable η varies as $0 \leqslant \eta \leqslant 1$. The coefficients of the matrices are defined as follows.

$$C_{ij} = \int_v \rho C_p N_i N_j dv$$

$$K_{ij} = \int_v \left\{k_x \frac{\partial N_i}{\partial x}\frac{\partial N_j}{\partial x} + k_x \frac{\partial N_i}{\partial x}\frac{\partial N_j}{\partial x}\right\} dv \tag{3.6}$$

$$H_{ij} = \int_S h N_i N_j ds, \quad \{q\}_i = \int_S q N_i ds$$

Where, N_i and N_j are shape functions [19]. In the one step method [18] for numerical time integration, the following assumptions are made for obtaining the solution to Equation (3.5) at time $(t+\eta_t)$:

$$T^{t+\eta\Delta t} = (1-\eta)T^t + \eta T^{t+\Delta t}$$
$$\frac{\partial T^{t+\eta\Delta t}}{\partial t} = \frac{(T^{t+\Delta t} - T^t)}{\Delta t} \qquad (3.7)$$

In order to obtaining the temperature distribution, above equations were solved using the Newton-Raphson iteration scheme.

2. Temperature models of EDM

A small cylindrical portion of the workpiece around spark is used as the domain (Fig. 3.3).

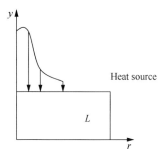

Fig. 3.3 Two-dimensional thermal model of the EDM

Heating of ceramics due to EDM (Gaussian distributed) is governed by the following transient heat conduction equation:

$$\rho C_s \frac{\partial T}{\partial t} = \frac{1}{r}\frac{\partial}{\partial r}\left(k_r \frac{\partial T}{\partial r}\right) + \frac{\partial}{\partial y}\left(k_y \frac{\partial T}{\partial y}\right) \qquad (3.8)$$

Where, r and y arc coordinate axes, C_s is the heat capacity.

The boundary conditions on the surface are

$$k_x \frac{\partial T}{\partial x} n_x + k_y \frac{\partial T}{\partial y} n_y + h(T - T_\infty) + q = 0 \qquad (3.9)$$

3. Superposition

Since temperature independent properties of the ceramics are considered, the combined temperature variation with time can also be obtained by the addition of known solutions of a simpler component problem [20]. According to this method different spatial temperature distributions with different initial conditions can be added to find final different distribution. Final temperature distribution of EDGSSDE is obtained by the superposition of the two temperatures as follows:

$$T_{TR} = T_0 + G_{TR} + EDM_{TR} \qquad (3.10)$$

Where, G_{TR} is the temperature rise due to grinding, EDM_{TR} is the temperature rise due to EDM, T_{TR} is the final temperature of EDGSSDE, T_0 is the initial temperature.

3.5.3 Residual stresses model

The steep temperature gradients occur during EDGSSDE result in nonuniformity in the local thermal expansion, which can cause residual stresses. The transient temperature distribution in the ceramics is used as input for the calculation of the residual stresses.

The relationship between stress-strain-temperature for a linearly elastic material is as follows:

$$d\{\sigma\} = [D]\left(d\{\varepsilon\} - d\{\varepsilon_1\} - d\{\varepsilon_1\}_T\right) \tag{3.11}$$

Where, $\{\sigma\}$ is the stress vector, $\{\varepsilon\}$ is the strain vector, $[D]$ is the elastic matrix, $\{\varepsilon_1\}$ is the original strain vector, $d\{\varepsilon_1\}_T$ is given by

$$d\{\varepsilon_1\}_T = \{\alpha\} + \frac{d[D]^{-1}}{dT}\{\sigma\}dT \tag{3.12}$$

Where, $\{\alpha\}$ is the differential strain vector.

The relationship between stress-strain-temperature for a plastic area is given by

$$d\{\sigma\} = [D]_{ep}\left(d\{\varepsilon\} - d\{\varepsilon_1\}_T + d\{\sigma_1\}_T\right) \tag{3.13}$$

Where, $[D]_{ep}$ is the elastic matrix for a plastic area, $\{\sigma_1\}$ is the original stress vector.

All strain increments for the ceramics being in the elastoplastic situation in the EDGSSDE can be expressed by the thermal elastoplastic equilibrium equation:

$$[K]\Delta\{U\} = \{\Delta R\} \tag{3.14}$$

Where, $[K]$ is the gross stiffness matrix, $\Delta\{U\}$ is the knot point displacement matrix, $\{\Delta R\}$ is the equal effect knot point load.

In the elastic area:

$$\Delta\{\sigma\} = [D]\left(\Delta\{\varepsilon\} - \Delta\{\varepsilon_1\}_T\right) \tag{3.15}$$

In the plastic area:

$$\Delta\{\sigma\} = [D]\left(\Delta\{\varepsilon\} - \Delta\{\varepsilon_1\}_T\right) + \Delta\{\sigma_1\}_T \tag{3.16}$$

Where,

$$\Delta\{\sigma_1\}_T = \frac{[D](\partial\sigma/\partial\{\sigma\})(\partial H/\partial T)\Delta T}{H_T + \{\partial\sigma/\partial\{\sigma\}\}^T[D](\partial\sigma/\partial\{\sigma\})} \quad (3.17)$$

$$\Delta\{\varepsilon_1\}_T = \left[\{\sigma\} + \left(\frac{d[D]^{-1}}{dT}\right)\{\sigma\}\right]\Delta T \quad (3.18)$$

Where, $\Delta\{\varepsilon_1\}_T$ is the strain coefficient of contact with temperature, $\Delta\{\sigma_1\}_T$ is the stress coefficient of contact with temperature, $[D]$ is the strain matrix, ΔT is the temperature increment, H_T is the submit coefficient.

3.5.4 Results

Two-dimensional isoparametric 8 node, 8 d.o.f. (degree of freedom) quadrilateral elements were used in this work. Moving heat is assumed to distribute over all elements at the top surface of the engineering ceramics. Gaussian distributed heat is assumed to be axisymmetric.

The temperature field of ceramics during grinding is shown in Fig. 3.4. Using present model top surface temperature is determined as 1148K. There is decrease in temperature with depth from top surface and increase in temperature towards bottom surface. The fact is realized that the rate of heat movement front into the workpiece exceeds the rate of down feed.

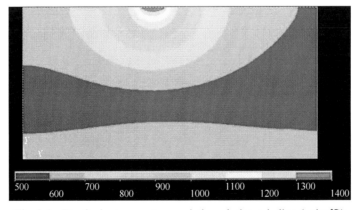

Fig. 3.4 Depthwise temperature variations during grinding (units: ℃)

The temperature field of ceramics during EDM is given by Fig. 3.5. With Gaussian heat flux, the top surface temperature is determined as 3007K. The trend of temperature variation along the radius is decreasing. High peak temperature appears, because the heat is only supplied during duration of the pulse and no heat supply for the interval.

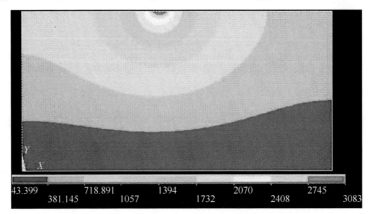

Fig. 3.5　Variation of temperature during EDM (units: ℃)

Depthwise temperature distribution in the engineering ceramics due to EDGSSDE is shown in Fig. 3.6. The temperature gradient variation is very high in a very thin surface layer. Beyond this depth temperature gradient variation is very small. This shows the temperature gradient variation is limited to a thin layer.

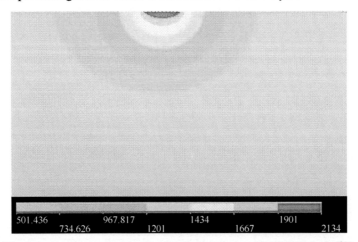

Fig. 3.6　Depthwise temperature distribution during EDGSSDE (units: ℃)

Fig. 3.7 shows the residual stresses calculated from the model. The graph shows the no-uniformity in residual stresses distribution.

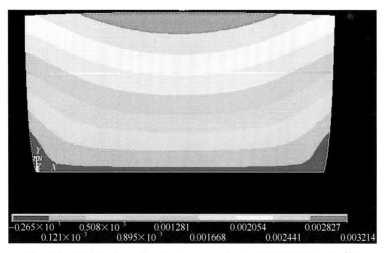

Fig. 3.7 Distribution of residual stresses during EDGSSDE (units: ℃)

3.6 Experiments

3.6.1 Experimental Conditions

Machining is performed on our self-developed insulating material electrical discharge grinding equipment with specially designed pulse power supply and electrode servo system. The EDGSSDE conditions of this research are presented in Table 3.1.

Table 3.1 The experimental conditions

Material of workpiece	Material of sheet electrode	Material of grinding wheel	Pulse power supply voltage	Pulse power supply current	Polarity	Working medium
Al_2O_3	Red copper	#45 steel	75~175V	30A max	Sheet electrode (−)	DX-1 emulsion for WEDM

Every workpiece is machined at the same feeding speed in the same time t. The masses of each workpiece are accurately measured with an electronic analytical balance before and after being machined. The MRR is calculated by Equation (3.19) and the SR is measured by a surface roughness tester.

$$\text{MRR} = \frac{m_0 - m_1}{t} \tag{3.19}$$

Where, m_0 is the mass of workpiece before being machined, m_1 is the mass of workpiece after being machined.

3.6.2 Results and analyses

1. Effects of pulse width

In order to study effects of pulse width on machining of insulating ceramics, a wide range of pulse width values (40μs, 50μs, 100μs, 200μs and 500μs) are selected. The other conditions of pulse power supply are set as follows.

(1) Pulse interval: 370μs.

(2) Peak voltage: 150V.

(3) Peak current: 25A.

Effects of pulse width on MRR and SR are showed in Fig. 3.8 and Fig.3.9. Pulse width in these two figures is in logarithmic scale.

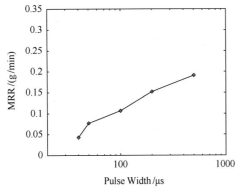
Fig. 3.8　Effect of pulse width on MRR

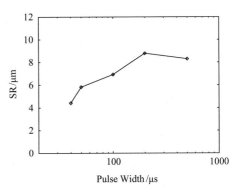
Fig. 3.9　Effect of pulse width on SR

Fig. 3.8 indicates that MRR almost grows linearly with logarithmic increasing pulse width. Because peak voltage does not vary, the breakdown delay time is the same, and the discharging time will grow with increasing pulse width. As Equation (3.20) [21] shows, the discharging energy increases with discharging time. In another view, if pulse width is long enough, discharge can happen more than one time during one pulse. So the rise of MRR with increasing pulse width can attribute to the two reasons above.

$$W = \int_0^{t_k} u(t)\,i(t)\,\mathrm{d}t \qquad (3.20)$$

Where, W is energy from the discharge, $u(t)$ is voltage between the discharging gap, $i(t)$ is current through the discharging gap, t_k is discharging time.

As a result of discharging sparks, there are many craters on the surface of workpiece. When pulse width is small, the size of craters increases fast with increasing pulse width. But the increasing speed will become lower because the transient discharging energy density does not vary with pulse width, and the max range affected by the discharging channel is certain. If pulse width is long enough and multi discharges happen in one pulse, the overlaps of craters can improve the surface roughness. So the curve in Fig. 3.9 can be well explained.

2. Effects of pulse interval

Setting pulse interval value to 400μs, 700μs, 1300μs, 2000μs and 2700μs, the workpieces are machined. The other conditions of pulse power supply are set as follows.

(1) Pulse width: 500μs.
(2) Peak voltage: 150V.
(3) Peak current: 25A.

Fig. 3.10 and Fig. 3.11 show how pulse interval affects MRR and SR according to the experimental results.

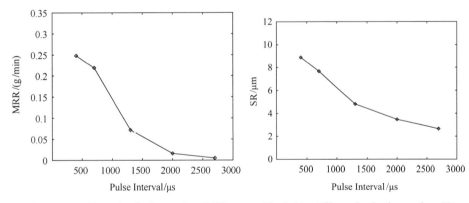

Fig. 3.10   Effect of pulse interval on MRR   Fig. 3.11   Effect of pulse interval on SR

It can be known that the discharging energy of every pulse is the same from conditions the machining used. So the sizes of craters are in a certain scale. Discharge frequency decreases when pulse interval grows, and the number as well as the density of craters will reduce. The MRR decreases with decreasing craters in a unit time, and a low density of craters means a low SR. The relationship can be revealed in Fig. 3.10 and Fig. 3.11.

3. Effects of peak voltage

Through setting peak voltage to different values as 75V, 100V, 125V, 150V and 175V, effects of peak voltage on MRR and SR are studied. The other parameters of pulse power supply are listed as follows.

(1) Pulse width: 500μs.
(2) Pulse interval: 370μs.
(3) Peak current: 25A.

The results are showed in Fig. 3.12 and Fig. 3.13.

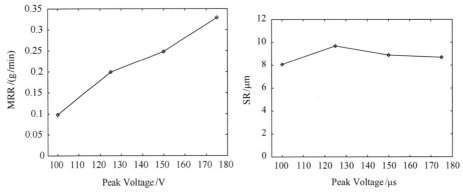

Fig. 3.12   Effect of peak voltage on MRR   Fig. 3.13   Effect of peak voltage on SR

Along with the increase of peak voltage, breakdown delay time decreases, and this make discharging time increase. As discussed before, because the pulse width selected is very long, discharge can happen more than one time, and the times increase while breakdown delay time decreases. Otherwise, the energy of every discharge increases when peak voltage rises. So it is certain that MRR increase with rising peak voltage.

The discharges get intenser and intenser while peak voltage rises, and the craters of workpiece surface get larger at the same time. But Fig. 3.13 indicates that the SR varies very little with increasing peak voltage. This can attribute to the overlaps of craters and grinding process of the wheel.

4. *Effects of peak current*

Peak currents of 5A, 10A, 15A, 20A and 25A were select to study their effect on MRR and SR of workpieces. The other pulse power supply parameters are as follows.

(1) Pulse width: 500μs.
(2) Pulse interval: 370μs.
(3) Peak voltage: 150V.

The effect of peak current on MRR is shown in Fig. 3.14. When peak current is 5A, the MRR is less than 0g/min. This is because the material eroded by sparks is less than the debris splashed on the workpiece from the two electrodes and working media.

As Equation (3.20) shows, discharging energy increases with electrical current. So the MRR increases with peak current.

Fig. 3.15 shows that SR rises while peak current increases from 5A to 20A, and then the value will not change much; it even descends in some measure due to grinding by the wheel.

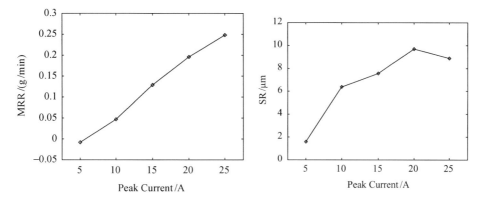

Fig. 3.14 Effect of peak current on MRR   Fig. 3.15 Effect of peak current on SR

3.7 Conclusions

A novel technique known as EDGSSDE for insulating engineering ceramics is pioneered in this chapter, the mechanism and characteristics are analyzed. This technique integrates the advantages of electrical discharge and mechanical grinding, achieves in-process dressing of grinding wheel, and makes the diamond grains always in knife-edge. The oxide layer on grinding wheel surface and the metamorphosed layer on workpiece surface debase the surface rigidity of workpiece, avoid micro-crack that is produced in traditional engineering grinding, and reduce the grinding force, which can improve the service life of grinding wheel, advance the surface quality and machine insulating engineering ceramics in ultra-precision.

A finite element-based model for the prediction of residual stresses generated by steep temperature gradients in engineering ceramics during EDGSSDE is presented. The usefulness of this work is that it can perform systematic virtual experiment for purely experimental investigation which is too expensive and sometimes difficult to perform. The present paper deals only with the residual stresses and the temperature fields of engineering ceramics during EDGSSDE. While EDGSSDE process is a combination of mechanical grinding and EDM, the complex phenomena including mechanical, thermal, electrical, chemical, etc is not well understood. The further research on this work is needed.

Experimental results show that effects of electrical parameters on the MRR and SR are very important in this insulating engineering ceramics electrical grinding method. Increase of pulse width, peak voltage, peak current or decrease of pulse interval can result in growth of MRR. Their effects on surface roughness are different in some degree. Usually, little parameters lead to a small SR, but peak voltage has little effect on surface roughness. As long as proper parameters are applied, higher MRR with acceptable surface quality can be obtained by this new electrical discharge grinding method.

References

[1] Sornakumar T. Advanced ceramic-ceramic composite tool materials for metal cutting applications. Key Engineering Materials, 1996, 114: 173-188.

[2] Yui A, Watanabe T, Yoshida Y. Effect of the static stiffness of a grinding machine on surface grinding accuracy of fine ceramics. Transactions of the Japan Society of Mechanical Engineers, Part C, 1987, 53: 2396-2399.

[3] Jahanmir S, Ives L K, Ruff A W, et al. Ceramic Machining: Assessment of Current Practice and Research Needed in the United States. NIST Special Publication, 1992:834.

[4] Fukuzawa Y, Mohri N. Machining characteristics of insulating ceramics by electrical discharge machine. Industrial Ceramic, 2001, 21: 187-189.

[5] Trueman C S, Huddleston J. Material removal by spalling during EDM of ceramics. Journal of the European Ceramic Society, 2000, 20: 1629-1635.

[6] Black I, Livingstone S A J, Chua K L. Laser beam machining (LBM) database for the cutting of ceramic tile. Journal of Materials Processing Technology, 1998, 84: 47-55.

[7] Jia Z X, Zhang J H, Xing A, et al. Combined machining of USM and EDM for advanced ceramics. Journal of Advanced Materials, 1995, 26: 16-20.

[8] Bandyopadhyay B P, Ohmori H, Takahashi I. Efficient and stable grinding of ceramics by electrolytic in-process dressing (ELID). Journal of Materials Processing Technology, 1997, 66: 18-24.

[9] Mohri N, Fukuzawa Y, Tani T, et al. Assisting electrode method for machining insulating ceramics. CIRP Annals- Manufacturing Technology, 1996, 45: 201-204.

[10] Hamdi H, Zahouani H, Bergheau J. Residual stresses computation in a grinding process. Journal of Materials Processing Technology, 2004, 147 (3): 277-285.

[11] Tian X L, Yu A B. A thermal elastoplastic model of the surface residual stress of ceramics grinding. Journal of Materials Processing Technology, 2002, 129 (1-3): 451-453.

[12] Das S, Klotz M, Klocke F et al. EDM simulation: Finite element-based calculation of deformation, microstructure and residual stresses. Journal of Materials Processing Technology,

2003, 142 (2): 434-451.

[13] Rebelo J C, Kornmeier M, Batista A C, et al. Residual stress after EDM-FEM study and measurement results. Materials Science Forum, 2002, 404-407: 159-164.

[14] Yadava V, Jain V K, Dixit P M, et al. Theoretical analysis of thermal stresses in electro-discharge diamond grinding. Machining Science and Technology, 2004, 8 (1): 119-140.

[15] Bifano T G, Dow T A, Scattergood RO. Ductile-regine grinding. A new technology for machining brittle materials. Journal of Engineering for Industry, Transactions of the ASME, 1991, 113: 184-189.

[16] Sreejith P S, Ngoi B K A. Material removal mechanisms in precision machining of new materials. International Journal of Machine Tools & Manufacture, 2001, 41: 1831-1843.

[17] Chen M J, Dong S, Li D, et al. Study on the influence factors of the surfaces quality in ultra-precision grinding machining of brittle materials. Chinese Journal of Mechanical Engineering, 2001, 37: 1-4, 10.

[18] Bathe K, Koshgoftaar M R. Finite element formulation and solution of nonlinear heat transfer. Journal of Nuclear Engineering and Design, 1979, 51: 389.

[19] Woodard P F, Chandrasekar S, Yang H T Y. Analysis of temperature and microstructure in the quenching of steel cylinders. Metallurgical Transactions, 1999, 30B: 815.

[20] Boley, B A, Weiner, J H. Theory of thermal stresses. New York: John Wiley and Sons, 1960.

[21] Li M H. Basic Theory of electrical discharge machining. Beijing: National Defence Industry Press, 1989.

Chapter 4 Electrical Discharge Milling of Weakly Conductive Engineering Ceramics

EDM is the extensively used non-conventional material removal process for machining engineering ceramics provided they are electrical conductive. However, the electrical resistivity of the popular engineering ceramics is higher, and there has been no research on the relationship between the EDM parameters and the electrical resistivity of the engineering ceramics that can be machined effectively by EDM. This chapter investigates the effects of the electrical resistivity and the EDM parameters on the EDM performance of ZnO/Al$_2$O$_3$ ceramic in terms of the machining efficiency and the quality. The experimental results shows that the electrical resistivity and the EDM parameters such as pulse duration, pulse interval, and peak current have the great influence on the machining efficiency and the quality during electrical discharge machining of ZnO/Al$_2$O$_3$ ceramic. Moreover, the electrical resistivity of the ZnO/Al$_2$O$_3$ ceramic, which could be effectively machined by EDM, increases with increases in pulse duration, peak current, and with decreasing the pulse interval, respectively. Furthermore, the ZnO/Al$_2$O$_3$ ceramic with the electrical resistivity up to 3410 Ω·cm could be effectively machined by EDM with the appropriate machining condition.

For weakly conductive engineering ceramics, this chapter employs a steel toothed wheel as the tool electrode to machine SiC ceramics with specific resistivity of 500 Ω·cm using electrical discharge milling. The process employs the pulse generator used in EDM, and uses a water-based emulsion as the machining fluid. It is able to effectively machine a large surface area on SiC ceramics with electrical resistivity of 500 Ω·cm, and effectively machine other advanced materials with high electrical resistivity such as polycrystalline diamond, and cubic born nitride. The effects of tool polarity, peak voltage, pulse duration, pulse interval, peak current, emulsion concentration, milling depth, rotational speed of the tool, tooth number of the tool, and tooth width of the tool on the process performance have been investigated.

4.1 Introduction

Engineering ceramics have outstanding properties such as a high strength-to-weight ratio, high hardness, high wear resistance and good chemical inertness at elevated temperature. Therefore, engineering ceramics are increasingly being used in the manufacturing industry, defense industry, aerospace, and other fields [1,2]. However, machining of these materials by using traditional machining methods is difficult due to high wear and corrosion resistance, hardness and brittleness properties of them [3-6].

Engineering ceramics are known as very difficult-to-machine materials [7]. The main factors that cause engineering ceramics to be difficult to machine are their high hardness, high strength and brittleness. Diamond grinding is one of the most commonly used techniques for engineering ceramic blanks shaping, but it is costly and inefficient. The high hardness of engineering ceramics induces higher grinding force and quick wear of diamond cutting edges [8-10].

EDM is one of the extensively used non-conventional material removal processes. Since there is no direct physical contact between the tool electrode and the workpiece, the process promises to be the effective and economical technique for machining conductive ceramics [11-14]. However, the electrical resistivity of the popular engineering ceramics is higher than that of metal, which makes machining engineering ceramics more difficult than machining metal using EDM. Recently, some researches about EDM performance of engineering ceramics have been done and the relevant published papers can be obtained.

Liu [15] investigated the electrical resistivity versus TiN content in TiN/Si_3N_4 composites. The results showed that a large number of TiN grains in the Si_3N_4 matrix could be connected to each other and formed the electrical conductive network leading to the increase in electrical conductivity. For TiN content higher than 30vol%, the electrical resistivity of the composites is approximately $0.01\Omega \cdot cm$, so that EDM can be used to machine this material. Qiu et al. [16] investigated the unidirectional conductivity of semiconductor crystals with the electrical resistivity of $2.1\Omega \cdot cm$ during EDM. The results showed that the semiconductor crystals with the electrical resistivity of $2.1\Omega \cdot cm$ could be directly machined by EDM, and the workpiece with positive polarity was more suitable for P-type semiconductor crystals during EDM. König et al. [17] machined ceramic materials with the EDM process, and found that the high removal rates as compared with traditional techniques for machining these materials could be obtained provided that their electrical conductivity values are of the order of 100Ω·cm.

It is also known that when the electrical resistivity of the engineering ceramics is low, the machinability of the engineering ceramics remains in the range of EDM, the discharge is stable, and the EDM performance, such as the MRR, EWR and SR is fine. As the electrical resistivity of the engineering ceramics increases, the discharge instability occurs, which will deteriorate the EDM performance of engineering ceramics. But so far there have been no research papers about the effect of the electrical resistivity on EDM performance, according to the authors. Moreover, there is no study about the relationship between the EDM parameters and the electrical resistivity of the engineering ceramics that can be machined effectively by EDM. Therefore, the effects of the electrical resistivity and machining parameters, such as pulse duration, pulse interval and peak current, on the EDM performance of engineering ceramics in terms of the machining efficiency and the quality, have been investigated in this chapter. The relevant results obtained in this chapter can promote the understanding of the electrical discharge machining mechanism of engineering ceramics, and enlarge the engineering ceramics range that can be machined by EDM.

4.2 Experimental procedures for EDM performance of engineering ceramics with different electrical resistivities

The effects of the electrical resistivity and machining parameters on the EDM performance are studied by using ZnO/Al_2O_3 ceramics with different Al_2O_3 contents, and their electrical resistivities are listed in Table 4.1. Although the electrical resistivity of ZnO ceramic is high, Al^{3+} can replace Zn^{2+} when Al_2O_3 is doped in ZnO. The Al^{3+} has one more positive carrier than Zn^{2+}, and there is one redundant electron which can move freely, hence, increase the electric conduction of ZnO. When the Al_2O_3 content of ZnO/Al_2O_3 ceramic increase, the carrier concentration induced by Al^{3+} increases, so the electrical resistivity decrease [18].

Table 4.1 The electrical resistivity of ZnO/Al_2O_3 ceramic with different Al_2O_3 content

Al_2O_3 content/wt%	0	0.01	0.05	0.1	0.5	1	1.5	2
Electrical resistivity/(Ω·cm)	324000	3410	2470	687	15	11.64	9	6.3

The experiment is performed in a self-developed EDM milling machine, as illustrated in Fig. 4.1. The electrode used is copper rods which is 4mm in diameter and a length of 10mm. The workpiece is ZnO/Al$_2$O$_3$ ceramic with different Al$_2$O$_3$ contents. The EDM performance is related to the efficiency which is determined in the EDM process by the MRR and by the EWR. The quality is determined by the microstructure and the SR, and the latter is characterized by using the arithmetic average roughness (R_a) value. During machining, the tool polarity is negative, the peak voltage is 150V, and the working fluid is composed of 10%(mass fraction) emulsified oil (DX-1, Nanjing Special Oil Factory, China) and 90% distilled water, which are mixed with a constant speed power-driver mixer (JJ-2, Jintan Medical Instrument Factory, China).

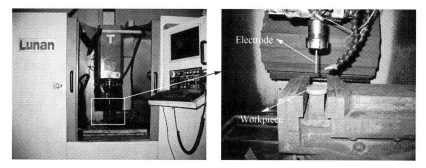

Fig. 4.1 EDM milling machine for machining ZnO/Al$_2$O$_3$ ceramic

The electrical resistivity of the ZnO/Al$_2$O$_3$ ceramic is measured with a pointer insulation resistance tester (PC40B, Shanghai Anbiao Electronics Corporation, China) and a digital ohmmeter (SD2002, Shanghai Qianfeng Electronic Instrument Corporation, China). The MRR and EWR are calculated according to Equation (4.1) and Equation (4.2), respectively, and they are obtained through measuring the weighing of the workpiece and the electrode before and after machining with an electronic balance (Sartorius BS224S, Germany). The SR was measured by a surface roughness tester (TR300, Qingdao Times Instrument Corporation, China). The microstructure of the workpiece surface was examined with an electron probe micro-analyzer (EPMA JEOL JXA-8230, Japan). All the observed specimens had been cleaned ultrasonically with the alcohol for twenty minutes and dried with a hot-air blower before the examination.

$$\text{MRR} = \frac{(m_1 - m_3)}{\rho_1 t} \tag{4.1}$$

$$\text{EWR} = \frac{\rho_1 (m_2 - m_4)}{\rho_2 (m_1 - m_3)} \times 100\% \tag{4.2}$$

Where, m_1, m_2 are the masses of engineering ceramic and tool electrode before machining, respectively, m_3, m_4 are the masses of engineering ceramic and tool electrode after machining, respectively, ρ_1, ρ_2 were the densities of engineering ceramic and tool electrode, respectively, t was the machining time.

4.3 Results and discussion of EDM performance of engineering ceramics with different electrical resistivities

4.3.1 Effect of the electrical resistivity and pulse duration on the process performance

The effect of the electrical resistivity and pulse duration on MRR is illustrated in Fig. 4.2, for a pulse interval of 50μs, and peak current of 50A. The MRR increases with the decrease of electrical resistivity. This phenomenon can be explained as follows: As the electrical resistivity decreases, the discharge channel is formed easily, the discharge delay time decreases, the pulse utilization ratio increases, as shown in Fig. 4.3, and the discharge energy released to the workpiece at the same time increases, so the MRR increases.

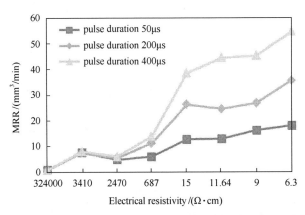

Fig. 4.2 Effect of the electrical resistivity and pulse duration on MRR

It can also be seen from Fig. 4.2 that the MRR increases slowly with an increase in pulse duration when the electrical resistivity is high, and the MRR increases rapidly with an increase in pulse duration when the electrical resistivity is low. There are many reasons causing this phenomenon. When the electrical resistivity is high, the discharge channel is formed difficultly with any pulse duration due to the low carrier concentration in the ZnO/Al$_2$O$_3$ ceramic, so the MRR is low, and it increases slowly. However, as the electrical resistivity is lower than 687Ω·cm, the discharge channel is formed easily, the discharge energy delivered into the machining zone within a single pulse increases with an increase in pulse duration, thermal erosion effects, such as melting and evaporation, are enhanced greatly, so the MRR increases rapidly with the increase in pulse duration.

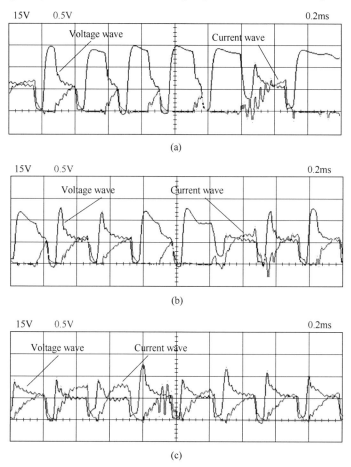

Fig. 4.3 Discharge waves of the ZnO/Al$_2$O$_3$ ceramics with different electrical resistivities. (a) the electrical resistivity of 3410Ω·cm, (b) the electrical resistivity of 687Ω·cm, (c) the electrical resistivity of 11.64Ω·cm

The effect of the electrical resistivity and pulse duration on EWR is shown in Fig. 4.4, for a pulse interval of 50μs, and peak current of 50A. The EWR initially increases with the decrease of electrical resistivity, and then decreases with the decrease of electrical resistivity. This phenomenon can be explained as follows: When the electrical resistivity is high, the discharge channel occurs difficultly, there are few discharges with any pulse duration, the tool electrode hits the ceramic easily due to the high electrical resistivity of the ceramic, and the ceramic removal is more than the electrode removal, so the EWR is low. When the electrical resistivity is lower than 687Ω·cm, the discharge is stable, the ceramic removal is high, a deposition layer can form on the electrode surface due to the decomposition of the dielectric and workpiece material attached to the tool electrode surface, and the electrode wear can be prevented by the protective effects of the deposition layer, so the EWR is low.

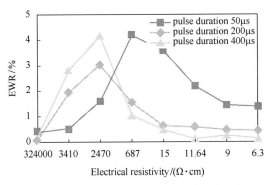

Fig. 4.4 Effect of the electrical resistivity and pulse duration on the EWR

It can also be seen from Fig. 4.4 that the EWR increases with the increase in pulse duration when the electrical resistivity is high, and the EWR decreases with the increase in pulse duration when the electrical resistivity is low. This is a complex phenomenon. When the electrical resistivity is high, the discharge channel is formed difficultly, but the discharge frequency with a long pulse duration is higher than that with a short pulse duration, the electrode material removal increment is more than the workpiece material removal increment due to the low melting point of the electrode with a long pulse duration, so the EWR increases with an increase in pulse duration. However, as the electrical resistivity is lower than 687Ω·cm, the discharge is stable, the ceramic removal is high, a deposition layer can form on the electrode surface to prevent the tool electrode wear, and the protective effect of the deposition layer increases with the increase in pulse duration [19]; therefore, the EWR decreases.

Fig. 4.5 shows the effect of the electrical resistivity and pulse duration on SR, for a pulse interval of 50μs, and peak current of 50A. With the pulse duration of 400μs, the SR initially decreases with the decrease of electrical resistivity, and then increases with the decrease of electrical resistivity; with the pulse duration of 200μs, the SR initially decreases with the decrease of electrical resistivity, and then changes slowly with the decrease of electrical resistivity; with the pulse duration of 50μs, the SR decreases with the decrease of electrical resistivity. There are many reasons causing this phenomenon. When the electrical resistivity is high, the discharge channel is formed difficultly, and there are few discharges in any pulse duration, the tool electrode hits the workpiece easily due to the high electrical resistivity of the workpiece, which deteriorates the workpiece surface, so the SR is high. However, as the electrical resistivity decreases, the discharge occurs relatively easily, the workpiece is removed by vaporization and melting, and the machined surface becomes smooth, so the SR decreases.

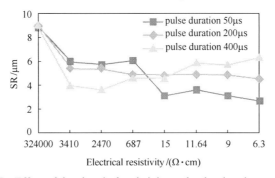

Fig. 4.5 Effect of the electrical resistivity and pulse duration on the SR

When the electrical resistivity is low and the pulse duration is set to 400μs, the discharge becomes violent, and the discharger crater is large due to the high single discharge energy, so the SR increases with the decrease in electrical resistivity. When the electrical resistivity is low and the pulse duration is set to 50μs, the discharge occurs easily, the discharger crater is small due to the low single discharge energy, and the discharger craters overlap easily, so the SR decreases with the decrease of electrical resistivity. When the electrical resistivity is low and the pulse duration is set to 200μs, the discharge condition and the crater size are between the above-mentioned two conditions, so the SR changes slowly with the decrease of electrical resistivity.

4.3.2 Effect of the electrical resistivity and pulse interval on the process performance

The effect of the electrical resistivity and pulse interval on the MRR is shown in Fig. 4.6, for a pulse duration of 200μs, and peak current of 50A. The MRR increases with the decrease in electrical resistivity, and the MRR with the pulse interval of 200μs is lower than that with the pulse interval of 20μs at the same electrical resistivity. This phenomenon can be explained as follows: As the electrical resistivity decreases, the discharge channel is formed easily, the discharge becomes violent, and the discharge energy released to the workpiece at the same time increases, so the MRR increases. Furthermore, the discharge frequency and the material removed in a unit time decrease with an increase in pulse interval, so the MRR with the pulse interval of 200μs is lower than that with the pulse interval of 20μs at the same electrical resistivity.

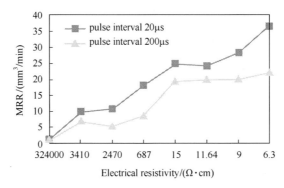

Fig. 4.6 Effect of the electrical resistivity and pulse interval on the MRR

The effect of the electrical resistivity and pulse interval on EWR is shown in Fig. 4.7, for a pulse duration of 200μs, and peak current of 50A. The EWR increases with the decrease of electrical resistivity. The reason for this is that the discharge is violent with the low electrical resistivity, and the electrode material removal increment is more than the workpiece material removal increment due to the high discharge energy and low melting point of the electrode. It can also be seen from Fig. 4.7 that the EWR with the pulse interval of 200μs is lower than that with the pulse interval of 20μs at the same electrical resistivity, especially when the electrical resistivity is low. This is because that the discharge frequency and the tool electrode removal decrease with an increase in pulse interval at the same electrical resistivity. Moreover, when the electrical resistivity is low, the electrode material removal with the pulse interval of 20μs is much higher than that with the pulse interval of 200μs due to the violent discharge. As the pulse interval decreases, the electrode material removal increment is more than the ceramic removal increment due to the low melting point of the electrode, so the EWR with the pulse interval of 200μs is much lower than that with the pulse interval of 20μs.

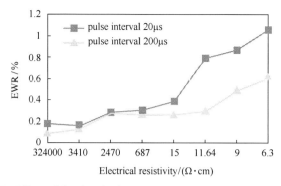

Fig. 4.7 Effect of the electrical resistivity and pulse interval on the EWR

Fig. 4.8 shows the effect of the electrical resistivity and pulse interval on SR, for a pulse duration of 200μs, and peak current of 50A. The SR decreases with the decrease of electrical resistivity, however, the SR increases with the decrease of electrical resistivity with the pulse interval of 20μs when the electrical resistivity is low. This is a complex phenomenon. When the electrical resistivity decreases, the discharge occurs easily, and the workpiece is removed by vaporization and melting, which makes the workpiece surface smooth, so the SR decreases with the decrease of electrical resistivity. However, when the electrical resistivity is lower than 11.64Ω·cm, the discharge becomes violent and the discharge crater is large and deep with the pulse interval of 20μs, so the SR increases with the decrease of electrical resistivity.

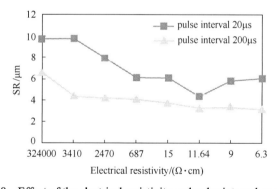

Fig. 4.8 Effect of the electrical resistivity and pulse interval on the SR

It can also be seen from Fig. 4.8 that the SR with the pulse interval of 20μs is higher than that with the pulse interval of 200μs at the same electrical resistivity. This phenomenon can be explained as follows: A longer pulse interval means more time for deionization of the dielectric, the discharge is stable, and the amount of craters generated by EDM is less; therefore, the SR decreases with an increase in pulse interval.

4.3.3 Effect of the electrical resistivity and peak current on the process performance

The effect of the electrical resistivity and peak current on the MRR is shown in Fig. 4.9, for a pulse duration of 200μs, and pulse interval of 50μs. The MRR increases with the decrease of electrical resistivity. There are many reasons causing this phenomenon. As the electrical resistivity decreases, the discharge channel is formed relatively easily, the discharge becomes violent, and the discharge energy released to the workpiece at the same time increases, so the MRR increases. It can also be seen from Fig. 4.9 that the MRR with the peak current of 10A is lower than that with the peak current of 30A at the same electrical resistivity, especially when the electrical resistivity is low. This is because the single discharge energy and the discharge crater volume increase with an increase in peak current. Moreover, when the electrical resistivity is lower than 687Ω·cm, the discharge becomes violent with the peak current of 30A, so the MRR is high.

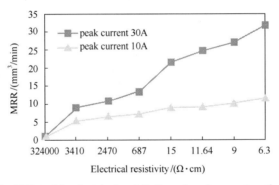

Fig. 4.9 Effect of the electrical resistivity and peak current on the MRR

The effect of the electrical resistivity and peak current on the EWR is shown in Fig. 4.10, for a pulse duration of 200μs, and pulse interval of 50μs. The EWR increases with the decrease of electrical resistivity. This phenomenon can be explained as follows: When the electrical resistivity decreases, the discharge channel is formed easily, the electrode material removal increases, and the electrode material removal increment is more than the workpiece material removal increment due to the low melting point of the electrode, so the EWR increases.

It can also be seen from Fig. 4.10 that when the electrical resistivity is high, the EWR with the peak current of 30A is higher than that with the peak current of 10A, however, when the electrical resistivity is low, the EWR with the peak current of 30A is lower than that with the peak current of 10A. There are many reasons causing this phenomenon. When the electrical resistivity is high, the discharge occurs difficultly, comparing with the peak current of 10A, the discharge with the peak current of 30A occurs relatively easily, which enhances the electrode material removal, and the electrode material removal is more easily than the ceramic removal, so the EWR with the peak current of 30A is higher. However, when the electrical resistivity is lower than 687Ω·cm, the discharge channel is formed easily, the discharge with the peak current of 30A is violent, and the workpiece material removal is greatly enhanced, the electrode material removal increment is lower than the ceramic removal increment, which can decrease the EWR, so the EWR with the peak current of 10A is higher than that with the peak current of 30A.

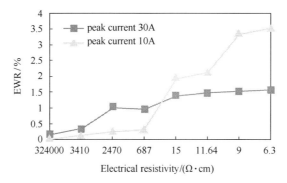

Fig. 4.10 Effect of the electrical resistivity and peak current on the EWR

Fig. 4.11 shows the effect of the electrical resistivity and peak current on SR, for a pulse duration of 200μs, and pulse interval of 50μs. The SR decreases with the decrease of electrical resistivity. This is because the discharge occurs easily, and the workpiece is removed by vaporization and melting when the electrical resistivity decreases, which makes the workpiece surface smooth.

It can also be seen from Fig. 4.11 that the SR with the peak current of 30A is lower than that with the peak current of 10A at the same electrical resistivity when the electrical resistivity is high, however, the SR with the peak current of 30A is higher than that with the peak current of 10A at the same electrical resistivity when the electrical resistivity is low. Many things can cause this phenomenon. When the electrical resistivity is high, there are few discharges on the workpiece surface, compared with the peak current of 10A, the discharge with the peak current of 30A occurs relatively easily, and the workpiece is removed by vaporization and melting, making the workpiece surface smooth, so the SR is low with the peak current of 30A. However, as the electrical resistivity is lower than 687Ω·cm, the discharge channel is formed easily, the discharge is violent, and the discharge crater is large and deep with the peak current of 30A, so the SR with the peak current of 30A is higher than that with the peak current of 10A.

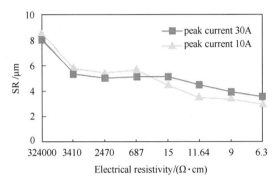

Fig. 4.11 Effect of the electrical resistivity and peak current on the SR

4.3.4 Microstructure character of ZnO/Al$_2$O$_3$ ceramic surface machined by EDM

The micrographs of ZnO/Al$_2$O$_3$ ceramic surfaces machined by EDM with different electrical resistivities are illustrated in Fig. 4.12, for a pulse duration of 400μs, pulse interval of 50μs, and peak current of 50A. The machined surface characteristics are obviously different with different electrical resistivities in the same machining condition.

Fig. 4.12(a) shows that the machined surface is rough with many particles when the electrical resistivity is 32400Ω·cm. This is because that the electrical resistivity is so high that there are few discharges during machining, and the electrode hits the workpiece surface easily, which deteriorates the machined surface.

Fig. 4.12(b)~Fig. 4.12(d) illustrate that there are some shallow craters, micropores and micro-cracks on the machined surfaces. The formation of the craters on these surfaces is due to sparks that form on the surface generating melting or possible evaporation. The formation of the micropores is due to entrapped gases that escape from the solidified material. The formation of the micro-cracks can be explained as follows: The rapid heating and cooling effects in EDM induce a high-temperature gradient within the heat affected area, and generate a significant stress within the machined surface. When the degree of the induced stress exceeds the maximum tensile strength of the ZnO/Al$_2$O$_3$ ceramic, cracking occurs on the machined surface. The surface characteristics such as micropores and micro-cracks become more pronounced as the electrical resistivity decreases. This is because the discharge channel is formed easily, the discharge becomes violent, and the discharge energy delivered to the machining gap increases with decreasing the electrical resistivity; therefore, the surface characteristics become more pronounced with decreasing the electrical resistivity.

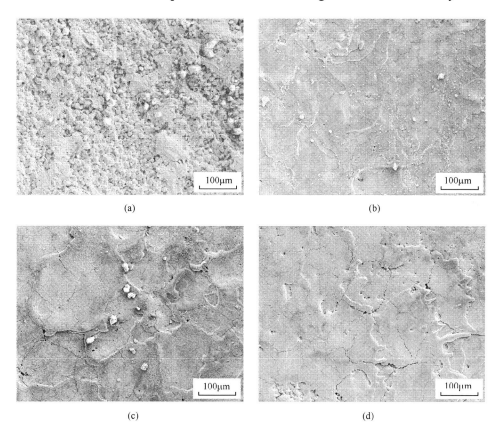

Fig. 4.12 Micrographs of machined ZnO/Al$_2$O$_3$ ceramic surfaces with different electrical resistivities. (a) The electrical resistivity of 32400Ω·cm, (b) The electrical resistivity of 2470Ω·cm, (c) The electrical resistivity of 687Ω·cm, (d) The electrical resistivity of 9Ω·cm

4.4 Principle for electric discharge milling of weakly conductive SiC ceramic

The principle for ED milling of SiC ceramic is shown in Fig. 4.13. The tool and the workpiece are connected to the positive and negative poles of the pulse generator respectively. The tool is a steel wheel with teeth and is mounted on to a rotary spindle, driven by an A. C. motor. The workpiece is SiC ceramic blank and is mounted on to an NC table. The machining fluid is a water-based emulsion.

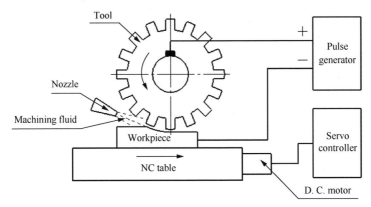

Fig. 4.13　Schematic illustration for ED milling of SiC ceramic

During machining, the tool rotates at a high speed; the SiC ceramic workpiece is fed towards the tool driven by a D. C. motor. As short-circuits or arcs are generated in ED milling, the workpiece is fed back by the D.C. motor. After short-circuits or arcs are cleared up, the workpiece is fed on again. The machining fluid is flushed into the gap between the tool and the workpiece with a nozzle. As the workpiece approaches the tool and the distance between the workpiece and a tooth of the tool reaches the discharge gap, electrical discharges are produced. A plasma channel grows during the pulse duration. A vapor bubble forms around this channel. The surrounding water-based emulsion restricts plasma growth, and makes the plasma energy densities rise to very high. The plasma temperature reaches nearly 40000K and plasma pressure can rise to 300MPa in EDM of conducting advanced ceramics [20]. The instantaneous high temperature and pressure plasma cause the SiC ceramic to be removed by ED milling. As the tooth has already left the workpiece and another tooth has not reached the discharge gap, electrical discharges are not produced. At this time machining fluid is flushed into the gap largely, the chips are flushed away easily and the workpiece is cooled quickly, which makes the processing stable. ED milling process uses steel for the electrode. This is because that during ED milling, the tool rotates at a high speed and the machining fluid is flushed into the discharge gap, the iron filings eroded for steel tool are very easy to be ejected away from the discharge gap. The unstable secondary discharges caused by the iron scraps can not be produced in ED milling. The iron and SiC ceramic filings produced by ED milling can abrade the tool electrode. ED milling process uses steel for the electrode, the resistance to abrasion of the steel electrode is higher than that of copper or graphite electrode. In addition, using steel as the tool material, the toothed electrode is manufactured easily and shows low cost.

4.5 Experiments and discussion for ED milling of weakly conductive SiC ceramic

In the following experiments, except the special instructions, the workpiece material is SiC ceramic with specific resistance of 500Ω·cm, the pulse duration is 50μs, the pulse interval is 400μs, the tool polarity is positive, the tool is a steel wheel with 40 teeth, the diameter of the tool is 180mm, the tool width is 4mm, the ratio of tooth thickness and circular pitch for the tool is 1 : 2, the rotational speed of the tool is 1100r/min, the work fluid is a water-based emulsion with the concentration of 5%. The weighing of the removed material or electrode wear is measured by an electronic analytical balance. The surface roughness is measured by a surface roughness tester. The microstructure of the SiC ceramic is observed by a scanning electron microscope, equipped by energy dispersive spectrometer analysis.

4.5.1 Effect of tool polarity on the process performance

The effect of tool polarity on the MRR, the wear ratio of the tool wear and ceramic removal, and the SR are given respectively in Fig. 4.14~Fig. 4.16.

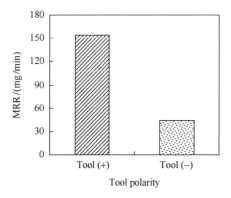

Fig. 4.14 Effect of tool polarity on the MRR

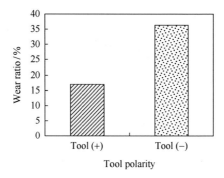

Fig. 4.15 Effect of tool polarity on wear ratio

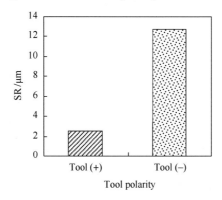

Fig. 4.16 Effect of tool polarity on the SR

The MRR with different tool polarities is given in Fig. 4.14. Under the same conditions, the MRR in positive tool polarity is 3.5 times that with negative tool polarity. The wear ratio with different tool polarities is given in Fig. 4.15. Under the same conditions, the wear ratio in negative tool polarity is 2 times that with positive tool polarity. These phenomena can be explained as follows: The longer pulse duration is used in the ED milling, the positive ions of discharge channel have enough time to be accelerated and the numbers of the positive ions arriving at the negative assisting electrode polarity close to the workpiece increase. Because the mass of the positive ions is much larger than that of electrons, the bombardment effect by the positive ions is stronger than that by electrons; therefore, MRR is high in positive tool polarity, the wear ratio is high in negative tool polarity.

Fig. 4.16 shows the influence of tool polarity on SR. Under the same conditions, the SR in negative tool polarity is five times that with tool positive polarity. This is because the stability of the electrical discharges becomes bad, and arcs are easily generated with negative tool polarity; therefore, the SR is high in negative tool polarity.

4.5.2　Effect of pulse duration on the process performance

The effect of pulse duration on the MRR and the SR is shown in Fig. 4.17 and Fig. 4.18, respectively. As shown in Fig. 4.17, the MRR initially increases rapidly with an increase of pulse duration and then decreases with an increase in pulse duration. There are many reasons causing this phenomenon. The longer the pulse duration with less current generated, the more thermal energy is lost owing to heat conduction; therefore, the MRR is low. However, with a very short pulse duration and a high current level, the thermal energy density is very high. At this time, the material is removed by vaporization. As the vaporization heat consumption is high, the MRR is low.

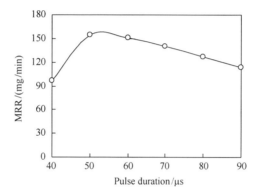

Fig. 4.17　Effect of pulse duration on the MRR

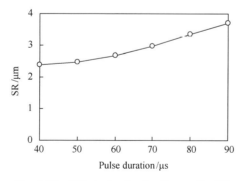

Fig. 4.18　Effect of pulse duration on the SR

Fig. 4.18 shows the influence of pulse duration on the SR. The SR increases with an increase in pulse duration. This phenomenon can be explained by the relationship between the SR and pulse duration as follows [21]:

$$R_{max} = K_R T_{on}^a I_p^b \tag{4.3}$$

Where, R_{max} is surface roughness (μm), T_{on} is pulse duration (μs), I_p is peak current (A), K_R is constant, a, b are proportional constants.

It can be seen from Equation (4.3) that under a certain peak current, the R_{max} increases with an increase in pulse duration, therefore, the SR increases.

4.5.3 Effect of pulse interval on the process performance

The effect of pulse interval on the MRR and the SR is shown respectively in Fig. 4.19 and Fig. 4.20, Fig. 4.19 shows the effect of pulse interval on the MRR. The MRR decreases with an increase in pulse interval. This phenomenon can be explained by the relationship between the material removed in a unit time and other parameters as follows [22].

$$V \propto KW\phi \frac{1}{t_{on} + t_{off}} \tag{4.4}$$

Where, V is material removal in a unit time, K is technological constant, W is the single pulse energy (J), ϕ is pulse utilization ratio, t_{on} is pulse duration (μs), t_{off} is pulse interval (μs).

It can be seen from Equation (4.4) that under the same conductions, the material removed in a unit time decreases with an increase in pulse interval, therefore, the MRR decreases.

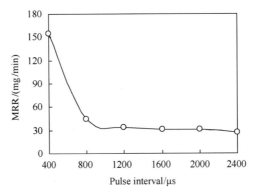

Fig. 4.19 Effect of pulse interval on the MRR

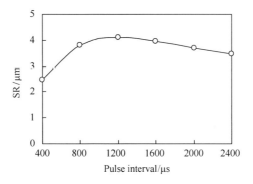

Fig. 4.20 Effect of pulse interval on the SR

As shown in Fig. 4.20, the SR initially increases rapidly with an increase in pulse interval and then decreases slowly with an increase in pulse interval. There are many reasons causing this phenomenon. With a longer pulse interval, there is more time for deionization of the dielectric, the breakdown voltage and the discharge explosion force increase, the crater size generated by a single pulse becomes larger and deeper; therefore, the SR increases with an increase in pulse interval. As the pulse interval is larger than 1200μs, the breakdown voltage and the discharge explosion force do not increase, the amount of crater generated by electrical discharge decreases; therefore, the SR decreases with an increase in pulse interval.

4.5.4 Effect of peak voltage on the process performance

The effect of peak voltage on the MRR and the SR is illustrated respectively in Fig. 4.21 and Fig. 4.22, Fig. 4.21 shows the relationship between the MRR and peak voltage. The figure indicates that the MRR rises with an increase in peak voltage. This phenomenon can be explained by relationship between the material removed by the single pulse and peak voltage as Equation (1.2). When the pulse duration and current-limit resistance are constant, the material removed by a single pulse increases with an increase in peak voltage; therefore, the MRR increases.

Effect of peak voltage on SR is shown in Fig. 4.22. The SR initially increases very slowly with an increase in the peak voltage and then increases rapidly with an increase in peak voltage. There are many reasons causing the phenomenon. The SiC ceramic has high electrical resistivity and high thermal decomposition point, as the peak voltage is less than 200V, the discharge current and the discharge explosion force is weak, the crater size generated by a single pulse is small and shallow; therefore, as the peak voltage is less than 200V, the SR increases slowly with an increase in peak voltage. As he peak voltage is higher than 200V, the discharge current and the discharge explosion force becomes stronger, the crater size generated by a single pulse becomes larger and deeper, and the material is even removed by explosion or flaking; therefore, the SR increases rapidly.

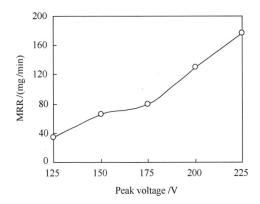

Fig. 4.21　Effect of peak voltage on the MRR

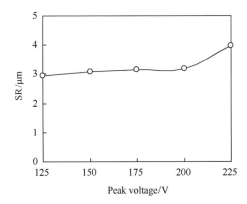

Fig. 4.22　Effect of peak voltage on the SR

4.5.5　Effect of peak current on the process performance

The effect of peak current on the MRR and the SR is shown in Fig. 4.23 and Fig. 4.24, respectively. As shown in Fig. 4.23, the MRR initially increases slowly with an increase of peak current and then increases rapidly with an increase in peak current. There are many reasons causing the phenomenon. The SiC ceramic has high electrical resistivity and high thermal decomposition point, as the peak current is less than 20A, the current density of electrical discharge is low, the SiC ceramic is difficult to be removed by the low current density of electrical discharge; therefore, as the peak current is less than 20A, the MRR increases slowly with an increase in peak current. As he peak current is higher than 20A, the electrical discharge becomes stronger and stability, the current density of electrical discharge is high; therefore, MRR rises rapidly with an increase in peak current.

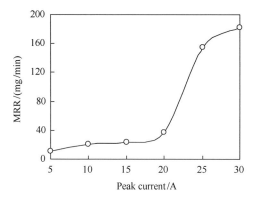

Fig. 4.23　Effect of peak current on material removal rate

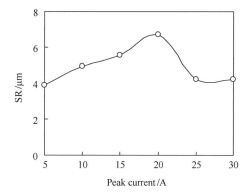

Fig. 4.24　Effect of peak current on surface roughness

Fig. 4.24 shows the influence of peak current on the SR. The SR initially increases with an increase in peak current then decreases with an increase in peak current. There are many reasons causing the phenomenon. The SiC ceramic has high electrical resistivity and high thermal decomposition point, as the peak current is less than 20 A, the concentrated discharges are easily produced, the crater size generated by a single pulse becomes larger and deeper, the crater size increases with increasing peak current; therefore, The SR increases with an increase in peak current. As the peak current is larger than 20A, the current density of electrical discharge is high, the concentrated discharges become very weak, the stability and uniformity of electrical discharges becomes stronger, the craters generated by electrical discharge are shallow and uniform distribution on the workpiece surface; therefore, SR is low.

4.5.6 Effect of emulsion concentration on the process performance

The effect of emulsion concentration on the MRR and the SR is illustrated in Fig. 4.25 and Fig. 4.26; the machining fluid is water and emulsion.

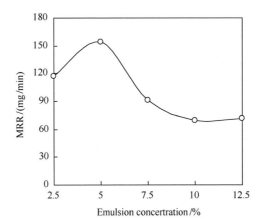

Fig. 4.25 Effect of emulsion concentration on the MRR

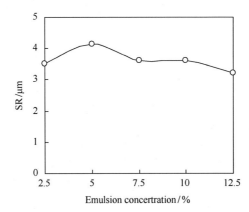

Fig. 4.26 Effect of emulsion concentration on the SR

As shown in Fig. 4.25, MRR initially increases with an increase in emulsion concentration and then decreases with an increase in emulsion concentration. There are many reasons causing the phenomenon. The dielectric strength, washing capability, density and viscosity of the machining fluid increase with an increase in emulsion concentration, pinch-effect and energy density of the discharge channel are enhanced, ejection effect of the eroded material increases; therefore, MRR rises. However, with a very high viscosity of the machining fluid the eroded materials are difficult to be flushed away, the electrical conductivity of the machining fluids between the tool and workpiece increases. Electrical discharges become weak; therefore, the MRR is low.

Fig. 4.26 shows the influence of emulsion concentration on SR. The SR initially increases with an increase of emulsion concentration, and then decreases with an increase in emulsion concentration. This is because the energy density of the discharge channel increases with an increase in emulsion concentration, the crater size generated by a single pulse becomes large; therefore, the SR rises with an increase in emulsion concentration. As the emulsion concentration is higher than 5%, the viscosity of the machining fluid increases largely. The eroded materials are difficult to be flushed away, the electrical conductivity of the machining fluids between the tool and workpiece increases. Electrical discharges become weak; therefore, the SR is low.

4.5.7 Effect of milling depth on the process performance

The effect of milling depth on the MRR is shown in Fig. 4.27. The MRR initially rises with an increase in milling depth, and then decreases with an increase in milling depth. There are many reasons causing this phenomenon. The shallower the milling depth with the higher discharge current density, the thermal energy density is very high. At this time, the material is removed by vaporization. As the vaporization heat consumption is high; therefore, the MRR is low. However, with a very deep milling depth the discharge current density is low, the SiC ceramic is difficult to be removed by the low current density of electrical discharge, the electrical discharge becomes unstable; therefore, as the milling depth is larger than 0.1mm, the MRR decreases with an increase in milling depth.

Fig. 4.28 shows the influence of milling depth on SR. The SR initially decreases with an increase in milling depth, and then increases with an increase in milling depth. The reason for this is that the shallower the milling depth with the higher discharge current density, the thermal energy density of the discharge channel is very high. The discharge explosion force increases largely; therefore, SR is high. However, with a very deep milling depth the discharge current density is low, the electrical discharge becomes unstable, arcs are easily generated; therefore, the SR is high.

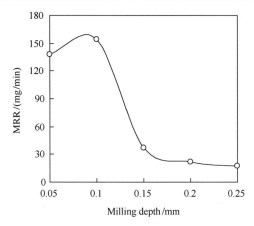

Fig. 4.27 Effect of milling depth on the MRR

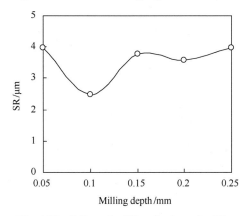

Fig. 4.28 Effect of milling depth on the SR

4.5.8 Effect of rotational speed on the process performance

As shown in Fig. 4.29, the MRR increases with an increase in rotational speed of the tool. There are many reasons causing the phenomena. The high rotational speed of the tool enhances the flow velocity of the work fluid flushing into the discharge gap, the migration velocity of discharge point on the tool surface increases. The cooling effect of the tool and the ejecting effect of the eroded materials enhance, the electrical discharges become strong and stable; therefore, the MRR increases with an increase in rotational speed of the tool.

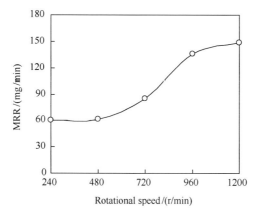

Fig. 4.29 Effect of rotational speed on the MRR

Fig. 4.30 shows the influence of rotational speed of the tool on SR. The SR initially increases with an increase in rotational speed of the tool, and then decreases with an increase in rotational speed of the tool. This is because the material removed by electric discharge becomes strong with an increase in rotational speed of the tool electrode, the crater size generated by electric discharge becomes large and deep; therefore, the SR rises. As the rotational speed of the tool is higher than 720r/min, the cooling effect of the tool and the ejecting effect of the eroded materials enhance greatly, the discharge points on the workpiece become dispersive and uniform. The craters generated by electrical discharges are shallow and uniform distribution on the workpiece surface; therefore, SR is low.

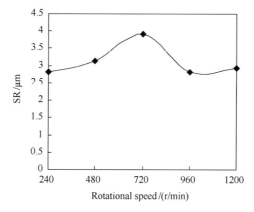

Fig. 4.30 Effect of rotational speed on the SR

4.5.9 Effect of tooth number on the process performance

The effect of tooth number of the tool on the MRR is shown in Fig. 4.31. The figure indicates that the MRR initially increases with an increase in tooth number and then decreases with an increase in tooth number. There are many reasons causing the phenomenon. The tooth thickness is constant, as the tooth number is little, the circular pitch increases. The electrical discharge times at a unite time is few, the electrical discharge times at a unite time increases with an increase in tooth number; therefore, the MRR increases with an increase in tooth number. However, as the tooth number is too more, the circular pitch of the tool is very little. The effect of mechanical deionization becomes weak, the electrical discharges become unstable, arcs are easily produced; therefore, as the tooth number is more than 40, the MRR decreases with an increase in tooth number.

As shown in Fig. 4.32, the SR initially decreases with an increase in tooth number of the tool, and then increases with an increase in tooth number of the tool. The reason causing this is that as the tooth number is little, the circular pitch is large, the cooling effect of the tool, the effect of mechanical deionization and the ejecting effect of the eroded materials become well, electrical discharge explosion force becomes strong, the crater size generated by electric discharge becomes large and deep; therefore, the SR is high. However, with a much more tooth number, the circular pitch of the tool is very little. The effect of mechanical deionization becomes weak, the electrical discharges become unstable, arcs are easily produced; therefore, the SR is high.

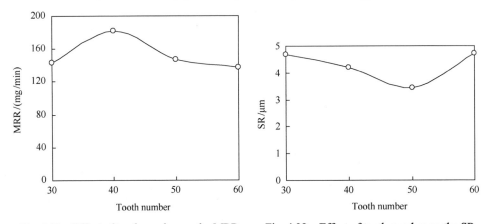

Fig. 4.31 Effect of tooth number on the MRR   Fig. 4.32 Effect of tooth number on the SR

4.5.10 Effect of tooth width on the process performance

Fig. 4.33 shows the effect of tooth width of the tool on the MRR. The MRR initially increases with an increase in tooth width of the tool, and then decreases with an increase in the tooth width. This is because the less tooth width with the higher current density, the thermal energy density is very high. At this time, the material is removed by vaporization. As the vaporization heat consumption is high, the MRR is low. However, with a very large tooth width, the current density and the thermal energy density are very low, electrical discharges become weak; therefore, the MRR is low.

The effect of tooth width of the tool on the SR is shown in Fig. 4.34. As the tooth width is less than 8, the SR initially decreases with an increase in the tooth width, and then increases with increasing tooth width, as the tooth width is larger than 8, the SR decreases with an increase in the tooth width. There are many reasons causing the phenomenon. As the tooth width is less than 8, the less tooth width with higher current density, the thermal energy density is very high. The electrical discharge explosion force becomes strong, the craters generated by electric discharges are large and deep; therefore, the SR is high. The current density and the thermal energy density decreases with an increase of tooth width, electrical discharges become weak; therefore, the SR decreases with increasing tooth width. However, as the tooth width is larger than 4, the current density decreases largely, electrical discharges become unstable and arcs are easily produced, therefore, the SR is high. As the tooth width is larger than 8, the current density decreases very largely, electrical discharges become very weak and electrical discharges are difficult to be produced, the craters generated by electric discharges are little and shallow; therefore, the SR is low.

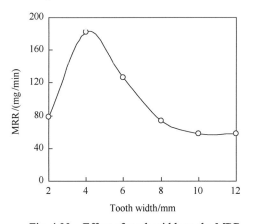
Fig. 4.33　Effect of tooth width on the MRR

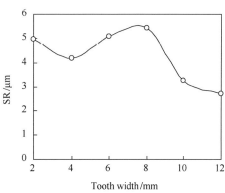
Fig. 4.34　Effect of tooth width on SR

4.5.11 Microstructure character of SiC surface machined by ED milling

The SEM micrograph of the SiC ceramic surface machined by ED milling is shown in Fig. 4.35. There are many craters and droplets left on the SiC ceramic surface machined by ED milling. These craters show irregular shape, different size and depth, because the micro electrical conductivity and micro organizational structure on the SiC ceramic surface are different. Some craters are deep and connecting each other, they are produced by electrical discharges along the electrical conductive network. The craters are mainly produced by the effects of oxidation and decomposition during ED milling of the SiC ceramic. A few of craters are produced by ED milling with melting and evaporation of metal additives in the SiC ceramic. The droplets are mainly formed by recast of the melting workpiece and tool materials. The material removal mechanisms in ED milling of SiC ceramics are that the most of materials are removed by the effects of oxidation and decomposition, a little of materials are removed by the effects of melting and evaporation.

Fig. 4.36 shows the cross-section of SiC ceramic machined by ED milling. The surface layer of the SiC ceramic workpiece has few pores. There are many pores in the

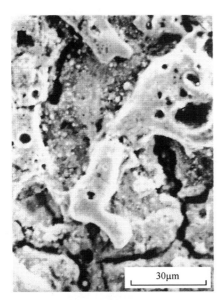
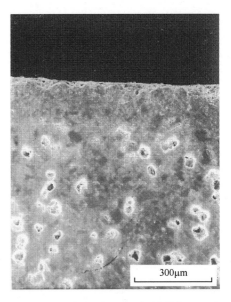

Fig. 4.35 SEM micrograph of SiC ceramic surface machined by ED milling

Fig. 4.36 SEM micrograph of SiC ceramic cross-section

base SiC ceramic, the crystal grains of the surface layer are smaller than that of the base material.

4.6 Conclusions

The electrical resistivity of the ZnO/Al$_2$O$_3$ ceramic, which can be effectively machined by EDM, increases with the increase in pulse duration, peak current, and with the decrease of pulse interval, respectively. Moreover, the ZnO/Al$_2$O$_3$ ceramic with the electrical resistivity up to 3410Ω·cm can be effectively machined by EDM with the appropriate machining condition.

Using a steel toothed wheel as tool electrode, the SiC ceramics with specific resistivity of 500 Ω·cm can be easily machined by ED milling. The process shows high MRR. Moreover, using a water-based emulsion as the machining fluid, harmful gas is not generated during ED milling of SiC ceramics, it shows good working environmental practice.

During machining, positive polarity for the tool electrode should be used. With the suitable pulse duration, pulse interval, peak voltage, and peak voltage, the high machining rate and good surface quality of machining the SiC ceramics with specific resistivity of 500 Ω·cm by ED milling can be easily obtained.

With a suitable emulsion concentration, milling depth, rotational speed of the tool, tooth number of the tool, and tooth width of the tool, the high machining rate and good surface quality for machining SiC ceramics with high electrical resistivity by ED milling can be easily obtained.

The crystal grains on the SiC ceramic surface machined by ED milling are smaller than that of the base material. The hardness and mechanical character of the SiC ceramic surface machined by ED milling are improved.

References

[1] Okada A. Automotive and industrial applications of structural ceramics in Japan. Journal of the European Ceramic Society, 2008, 28(5): 1097-1104.
[2] Liang X H, Lin B, Han X S, et al. Fractal analysis of engineering ceramics ground surface. Applied Surface Science, 2012, 258(17): 6406-6415.
[3] Feng J, Chen P, Ni J. Prediction of grinding force in microgrinding of ceramic materials by cohesive zone-based finite element method. International Journal of Advanced Manufacturing

Technology, 2013, 68(5-8): 1039-1053.

[4] Lalchhuanvela H, Doloi B, Bhattacharyya B. Analysis on profile accuracy for ultrasonic machining of alumina ceramics. International Journal of Advanced Manufacturing Technology, 2013, 67(5-8): 1683-1691.

[5] Chen S, Lin B, Han X, et al. Automated inspection of engineering ceramic grinding surface damage based on image recognition. International Journal of Advanced Manufacturing Technology, 2013, 66(1-4): 431-443.

[6] Yue Z, Huang C, Zhu H, et al. Optimization of machining parameters in the abrasive waterjet turning of alumina ceramic based on the response surface methodology. International Journal of Advanced Manufacturing Technology, 2014, 71(9-12): 2107-2114.

[7] Luis C J, Puertas I, Villa G. Material removal rate and electrode wear study on the EDM of silicon carbide. Journal of Materials Processing Technology, 2005, 164/165: 889-896.

[8] Agarwal S, Venkateswara R P. A new surface roughness prediction model for ceramic grinding. Proceedings of the Institution of Mechanical Engineers, Part B: Journal of Engineering Manufacture, 2005, 219(11): 811-821.

[9] Li X P, Liu Y H, Ji R J, et al. Non-traditional machining technique for insulating engineering ceramics. Electromachining & Mould, 2006, (5):6-9.

[10] Gopal A V, Rao P V. Modeling of grinding of silicon carbide with diamond wheels. Mineral Processing and Extractive Metallurgy Review, 2002, 23(1): 51-63.

[11] Rao T, Krishna A. Selection of optimal process parameters in WEDM while machining Al7075/SiC$_p$ metal matrix composites. International Journal of Advanced Manufacturing Technology, 2014, 73(1-4): 299-314.

[12] Wei C, Zhao L, Hu D, Ni J. Electrical discharge machining of ceramic matrix composites with ceramic fiber reinforcements. International Journal of Advanced Manufacturing Technology, 2013, 64(1-4): 187-194.

[13] Ignacio P, Carmelo J L. Optimization of EDM conditions in the manufacturing process of B_4C and WC-Co conductive ceramics. International Journal of Advanced Manufacturing Technology, 2012, 59(5-8): 575-582.

[14] Patel K M, Pulak M P, Venkateswara R P. Optimisation of process parameters for multi-performance characteristics in EDM of Al_2O_3 ceramic composite. International Journal of Advanced Manufacturing Technology, 2010, 47(9-12): 1137-1147.

[15] Liu C C. Microstructure and tool electrode erosion in EDMed of TiN/Si_3N_4 composites. Materials Science and Engineering: A, 2003, 363(1/2): 221-227.

[16] Qiu M, Liu Z, Tian Z, et al. Study of unidirectional conductivity on the electrical discharge machining of semiconductor crystals. Precision Engineering, 2013, 37(4): 902-907.

[17] König W, Dauw D F, Levy G, et al. EDM-future steps towards the machining of ceramics. CIRP Annals - Manufacturing Technology, 1988, 37(2): 623-631.

[18] Wang Z, Peng C, Wang R, et al. Precipitation of Al-doped-ZnO(AZO) ceramic targets and determination of its resistance properties. The Chinese Journal of Nonferrous Metals, 2013, 23(12): 3341-3347.

[19] Ji R J, Liu Y H, Zhang Y Z, et al. High-speed end electric discharge milling of silicon carbide ceramics. Materials and Manufacturing Processes, 2011, 26(8): 1050-1058.
[20] Petrofes N F, Gadalla A M. Electrical discharge machining of advanced ceramics. Ceramic Bull, 1988, 67(6):1048-1052.
[21] Cao F G. Electron discharge machining. Beijing: Chemical Industry Press, 2005.
[22] Liu J C, Zhao J Q. Non-conventional machining. Beijing: Machine Industry Press, 2005.

Chapter 5 End Electric Discharge Milling of Weakly Conductive Engineering Ceramics

Weakly conductive engineering ceramics, such as SiC ceramics, have found a variety of engineering applications due to their superior properties. However, the manufacture of weakly conductive engineering ceramics is not an efficient process. A new process of machining weakly conductive engineering ceramics using end ED milling is presented in this chapter. End ED milling employs a turntable with several small cylindrical rods as the tool electrode, and uses a water-based emulsion as the machining fluid. This process is able to effectively machine a large surface area on weakly conductive engineering ceramics. The machining principle and characteristics of the technique are introduced. The effects of tool polarity, pulse duration, pulse interval, peak voltage, peak current, emulsion concentration, emulsion flux, milling depth, copper electrode number, and copper electrode diameter on the process performance such as the MRR, EWR, and SR have been investigated with Taguchi experimental design method. Analysis of variance (ANOVA) and F-test are used to indicate the significant machining parameters affecting the machining characteristics. Furthermore, mathematical models relating to the machining characteristics are established with the stepwise regression method. Confirmation experiment results show that the machining performance can be improved significantly by using Taguchi experimental design method, and the developed mathematical models are appropriate for the effective machining of SiC ceramic.

The surface microstructures machined by the new process are examined with SEM, EDS, and XRD. The results show that the SiC ceramic is removed by melting, evaporation and thermal spalling, the material from the tool electrode can transfer to the workpiece, and a combination reaction takes place during end electric discharge milling of the SiC ceramic.

5.1 Introduction

Weakly conductive engineering ceramics, such as SiC ceramics, have intrinsic characteristics such as high hardness, high strength, good chemical inertness and high wear resistance, that makes them promising candidates for high temperature structural and wear-resistance materials [1-3]. The applications of SiC ceramics include dies, cutting tools, seal rings, valve seats, bearing parts, and a variety of engine parts etc [4-6]. However, the brittle nature and low fracture toughness of SiC ceramics make machining difficult and costly by conventional machining methods [7,8].

Diamond grinding and diamond turning are the major machining methods for SiC ceramic. Agarwal and Rao [9] ground SiC ceramic with a diamond wheel and investigated the grinding characteristics, surface integrity and material removal mechanisms. Yin et al. [10] studied the plastic deformation, fracture damage and material removal mechanisms during spherical grinding of polycrystalline silicon carbide with a diamond tool. Gopal and Rao [11] developed a new chip-thickness model for the performance assessment of silicon carbide grinding by incorporating elastic properties of the diamond grinding wheel and the workpiece. Zhang et al. [12] studied the precision machinability of reaction-bonded silicon carbide by single-point diamond turning and investigated the influence of the depth of the cut and tool feed rate on surface roughness and cutting force. The above-mentioned studies have many advantages; however, diamond grinding and diamond turning give rise to difficulties associated with the high cost of diamond tools, large consumption of diamonds and laborious processing due to the high hardness and brittleness of the silicon carbide ceramic.

EDM is a non-traditional manufacturing process and uses thermal energy induced by an electrical discharge spark in material removal. Therefore, it can be applied to the machining process for all of the conductive materials, regardless of their hardness. Moreover, EDM alleviates the concerns about the problems in conventional machining processes, such as tool breakage, vibration, and deflections caused by mechanical contact forces because this process is a noncontact material removal process. Hence, silicon carbide ceramic, which is a difficult-to-machine material, can be machined effectively by EDM [13-15]. However, conventional EDM such as die-sinking electrical discharge machining, the WEDM and EDG shows low efficiency when machining a large surface area on SiC ceramic [16-18].

EDM milling uses the rotating solid cylindrical electrode or tubular electrode to contour along a path. The process has been developed and demonstrated to be beneficial for its simplicity of electrode geometry [19,20]. Recently, some researches have been made to improve the efficiency with EDM milling. The results show that the material removal rate can be improved proportionally, but it still can't meet the demand of modern industrial applications, and the machined surface is poor [21-23].

A novel high speed end ED milling method using a turntable with several small cylindrical rods as the tool electrode has been proposed in this chapter. The process employs the pulse generator used in EDM, and uses a water-based emulsion as the machining fluid. It is able to effectively machine a large surface area on SiC ceramic. The effects of tool polarity, pulse duration, pulse interval, peak voltage, peak current, emulsion concentration, emulsion flux, milling depth, copper electrode number, and copper electrode diameter on the process performance such as the MRR, EWR, and SR have been investigated with Taguchi experimental design method. Moreover, the surface microstructures machined by the new process are examined with SEM, EDS, and XRD.

5.2 Principle and characteristics for end ED milling of weakly conductive SiC ceramic

5.2.1 Principle for end ED milling of SiC ceramic

The principle for end ED milling of weakly conductive SiC ceramic is shown in Fig. 5.1. The tool and the workpiece are connected to the negative and positive poles of the pulse generator, respectively. The tool is a turntable with several small cylindrical rods rotating rapidly around its axis. The tool is mounted onto a rotary spindle, driven by an A.C. motor. The workpiece is SiC ceramic blank and is mounted on to a NC table. The machining fluid is a water-based emulsion.

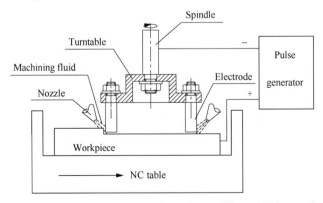

Fig. 5.1 Schematic illustration for end ED milling of SiC ceramic

During machining, the tool rotates at a high speed; the machining fluid is flushed into the gap between the tool and the workpiece with several nozzles, the SiC ceramic workpiece is fed towards the tool with NC table. As the workpiece approaches the tool and the distance between the workpiece and the electrode reaches the discharge gap, electrical discharges are produced. A plasma channel with high temperature and high pressure grows during the pulse duration. The instantaneous high temperature and pressure plasma cause SiC ceramic to be removed by electrical discharge milling. As short circuits or arcs are generated in end ED milling, the workpiece is fed back by the NC table. After short circuits or arcs are cleared up, the workpiece is fed on again. In addition, the rapid rotation of the tool and layer-by-layer machining cause a rapid stabilization of the cylindrical electrodes shape. Wear will cause the cylindrical electrodes to become shorter, but their shapes remain virtually unchanged.

5.2.2 Characteristics for end ED milling of SiC ceramic

The advantages of end ED milling SiC ceramic are listed below.

(1) A large diameter turntable with several cylindrical copper rods is used as the tool (Fig. 5.2), so a large surface area can be easily machined by end ED milling. The process shows high MRR. Fig. 5.3 is a photograph of the SiC ceramic workpiece machined by end ED milling.

Fig. 5.2 Tool photograph used in end ED milling

(2) Using cylindrical copper rods as the electrode (Fig. 5.4), the electrode is manufactured easily and shows low cost. Furthermore, the cylindrical copper rods are fixed and replaced easily during end ED milling, which can eliminate the need for monolithic tool electrode with high cost.

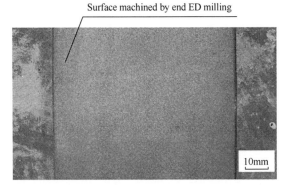

Fig. 5.3 Photograph of SiC ceramic workpiece machined by end ED milling

Fig. 5.4 Photograph of cylindrical copper rods used in end ED milling

(3) The cylindrical electrodes are fixed on the turntable separately, the working fluid is flushed into the gap, and the chips are flushed away easily. Therefore, end ED milling improves the stability of the electrical discharges.

(4) A water-based emulsion is used as the machining fluid, so harmful gas is not generated during the end ED milling, and the equipment is not corroded.

(5) Rough machining, semi-finish machining and finish machining can be obtained on the same machine by adjusting the discharge parameters when large surface is machined on the SiC ceramic workpiece.

(6) With some specific motion controls, end ED milling can machine complex cavities on the SiC ceramic workpiece efficiently.

5.3 Experiments

5.3.1 Experimental procedures

In the following experiments, the workpiece material is SiC ceramic with the electrical resistivity of 500Ω·cm, the tool is a turntable with uniform-distributed cylindrical copper electrodes in the circumference. The diameter of the turntable is 90mm, the rotational speed of the spindle is 3000r/min. The machining fluid is a water-based emulsion composed of emulsified oil and distilled water, which are mixed with a constant speed power-driver mixer. The velocity of the machining fluid over the workpiece is 1.67m/s. The material removal rate and electrode wear ratio are shown in Equation (5.1) and Equation (5.2), respectively, and they are obtained through measuring the dimensions of the workpiece and the electrode before and after machining with a dial indicator and a vernier caliper. The SR is measured by a surface roughness tester, and the SR is obtained by averaging five measurements made stochastically at different positions on the machined surface. The microstructure of the workpiece surface is examined with SEM, EDS, and XRD. All the observed specimens have been cleaned ultrasonically and dried with a hot-air blower before the examination.

5.3.2 Experimental design

To evaluate the effects of machining parameters on machining characteristics, and to identify the machining characteristics under the optimal machining parameters, a specially designed experimental procedure is required. Taguchi method in general, provides a significant reduction in the size of experiments, thereby speeds up the experimental process [24-26]. In this chapter, the experimental design employed a L_{25} orthogonal array based on the Taguchi method. The L_{25} orthogonal array have six columns and twenty-five rows, so it have twenty-four degrees of freedom to treat six parameters with five levels. Thus, six machining parameters are assigned to the columns and the rows specifiy twenty-five experiments with various combination levels of the machining parameters. In this investigation, only four machining parameters such as open voltage (U), discharge current (I), pulse duration (t_i) and pulse interval (t_o) are chosen to explore their impacts on the machining characteristics in terms of MRR, EWR and SR. The orthogonality is remained, even if two columns of the array remain empty. Table 5.1 presents the machining parameters and their levels used in this investigation.

Table 5.1 Machining parameters and their levels

Factors	Levels				
	1	2	3	4	5
U/V	90	110	130	150	170
I/A	15	30	45	60	75
t_i/μs	400	120	90	50	40
t_o/μs	300	600	1200	1800	2500

5.3.3 Analysis and discussion of experimental results

Experiments are replicated three times and randomized to minimize the bias from both between experiments and within experiment error. A better feel for the relative effect of the different machining parameters on the machining characteristics is obtained by ANOVA. The relative importance of the machining parameters with respect to the MRR, EWR and SR is investigated to determine more accurately the optimum combinations of the machining parameters by using ANOVA. Statistically, F-test provides a decision at some confidence level as to whether these estimates are significantly different. Larger F-value indicates that the variation of the process parameter makes a big change on the machining characteristics. F-values of the machining parameters are compared with the appropriate confidence table. When the F-value of the machining parameter is bigger than F_{α, n_1, n_2} value of the confidence table, where α is risk, n_1 and n_2 are degrees of freedom associated with numerator and denominator, the contribution of the machining parameter is defined as significant. During this investigation, there are three categories of significant machining parameters: ① more significant machining parameters (α is 0.001); ② significant machining parameters (α is 0.01); ③ less significant machining parameters (α is 0.1). F_{α, n_1, n_2} is quoted from the "Tables for Statisticians" [27].

To obtain applicable and practical predictive quantitative relationships, it is necessary to model the machining parameters and the machining characteristics. These models will be of great use during optimization of the SiC ceramic machining. In this investigation, a series of mathematical models are obtained by using the stepwise regression method with a commercially available mathematical software SAS (version 8.0).

5.4 Results and discussion of the single factor experiment during end ED milling

5.4.1 Effect of tool polarity on the process performance

Tool polarity is a primary factor that affects the process performance such as the MRR, the EWR, and the SR. The effect of tool polarity on the process performance is given in Fig. 5.5.

The MRR with different tool polarities is given in Fig. 5.5(a). Under the same conditions, the MRR in negative tool polarity is 1.6~1.9 times higher than that with positive tool polarity. The EWR with different tool polarities is given in Fig. 5.5(b). Under the same conditions, the EWR in positive tool polarity is 3.6~4.4 times higher than that with negative tool polarity. The phenomenon can be explained as follows. The tool rotates at a high speed during machining, the discharge point transfer velocity between the copper electrode and workpiece is very high, and the continuance time of the discharge point between the certain point of the copper electrode and the certain point of the workpiece is very short. Because the mass of the electrons is much smaller than that of positive ions, and they can be accelerated quickly during a short time, the bombardment effect by electrons is stronger than that by positive ions; therefore, the MRR is high in negative tool polarity, the EWR is high in positive tool polarity.

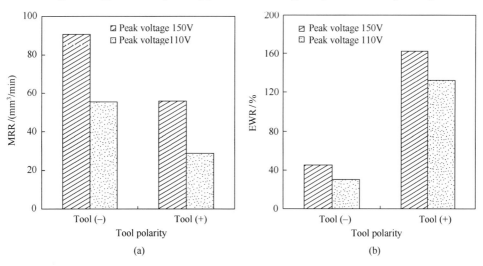

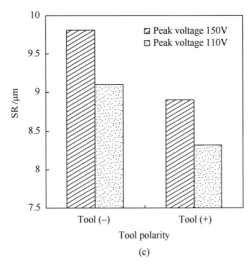

Fig. 5.5 Effect of tool polarity on the process performance. (a) Effect of tool polarity on the MRR, (b) Effect of tool polarity on the EWR, (c) Effect of tool polarity on the SR

Fig. 5.5(c) shows the effect of tool polarity on SR. Under the same conditions, the SR in negative tool polarity is 1.1 times higher than that with positive tool polarity. Because the bombardment effect by positive ions is weaker than that by electrons, the craters on the workpiece surface produced by positive ions are shallow; therefore, the SR is low in positive tool polarity.

5.4.2 Effect of pulse duration on the process performance

The effect of pulse duration on MRR, EWR and SR is shown in Fig. 5.6(a), Fig. 5.6(b) and Fig. 5.6(c), respectively. As shown in Fig. 5.6(a), the MRR increases gradually with an increase in pulse duration. This is because discharge energy delivered into the machining zone within a single pulse increases with an increase in pulse duration, thermal erosion effects, such as vaporization and melting on the machined surface, are enhanced, the MRR increases with an increase in pulse duration.

Fig. 5.6(b) shows the effect of pulse duration on EWR. The EWR decreases with an increase in pulse duration. This phenomenon can be explained as follows: During EDM, a deposition layer can form on the electrode surface due to the decomposition of the dielectric and workpiece material attached to the tool electrode surface, and the tool electrode wear can be prevented by the protective effects of the deposition layer [28,29]. As the pulse duration increases, the discharge energy delivered to the machining gap increases, the dielectric and workpiece are heated for more time, the released carbon decomposed from the dielectric is easily attached to the electrode surface, the deposition effect is enhanced, the tool electrode wear decreases; therefore, the EWR decreases.

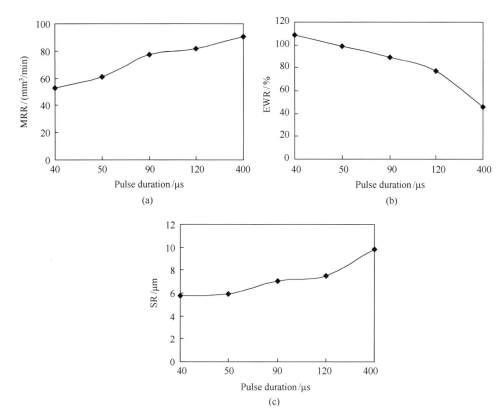

Fig. 5.6 Effect of pulse duration on the process performance. (a) Effect of pulse duration on the MRR, (b) Effect of pulse duration on the EWR, (c) Effect of pulse duration on the SR

The effect of pulse duration on SR is shown in Fig. 5.6(c). The SR increases with an increase in pulse duration. This phenomenon can be explained by the relationship between the SR and pulse duration as Equation (4.3), under a certain peak current, the R_{max} increases with an increase in pulse duration; therefore, the SR increases.

5.4.3　Effect of pulse interval on the process performance

The effect of pulse interval on MRR, EWR and SR is illustrated in Fig. 5.7(a), Fig. 5.7(b) and Fig. 5.7(c), respectively.

Fig. 5.7(a) shows the effect of pulse interval on MRR. The MRR decreases with an increase in pulse interval. This phenomenon can be explained by the relationship between the material removed in a unit time and the other parameters as Equation (4.4), under the same conditions, the material removed in a unit time decreases with an increase in pulse interval, therefore, the MRR decreases.

Fig. 5.7(b) shows the effect of pulse interval on EWR. The EWR increases with an increase in pulse interval. This phenomenon can be explained as follows: The time for deionization of the dielectric increases with an increase in pulse interval, the discharge energy delivered to the machining gap decreases in a unit time, the released carbon decomposed from the dielectric decreases, the deposition effect weakens, the tool wear compensation decreases; therefore, the EWR increases.

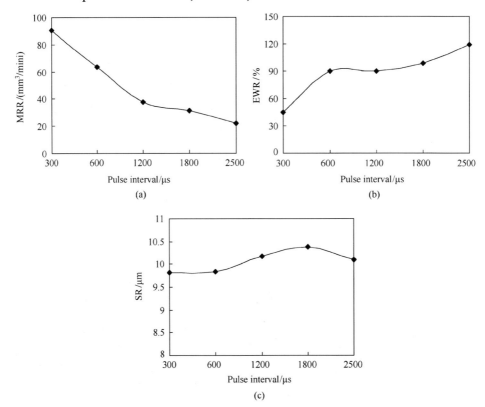

Fig. 5.7 Effect of pulse interval on the process performance. (a) Effect of pulse interval on the MRR, (b) Effect of pulse interval on the EWR, (c) Effect of pulse interval on the SR

As shown in Fig. 5.7(c), the SR initially increases with an increase in pulse interval, and then decreases with an increase in pulse interval. Many reasons cause this phenomenon. A longer pulse interval means more time for deionization of the dielectric, the breakdown voltage and the discharge explosion force increase, the crater size generated by a single pulse becomes larger and deeper; therefore, the SR increases with an increase in pulse interval. As the pulse interval is larger than 1800μs, the breakdown voltage and the discharge explosion force do not increase, the amount of crater generated by electrical discharge decreases; therefore, the SR decreases with an increase in pulse interval.

5.4.4 Effect of peak voltage on the process performance

The effect of peak voltage on the MRR, EWR and SR is shown in Fig. 5.8(a), Fig. 5.8(b) and Fig. 5.8(c), respectively. Fig. 5.8(a) shows the relationship between the MRR and peak voltage. The MRR increases with an increase in peak voltage. This phenomenon can be explained by the relationship between the material removed by the single pulse and peak voltage as Equation (1.2), under the constant pulse duration and current-limit resistance, the material removed by a single pulse increases with an increase in peak voltage; therefore, the MRR increases.

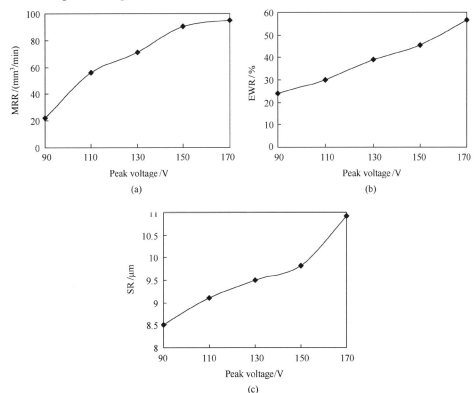

Fig. 5.8 Effect of peak voltage on the process performance. (a) Effect of peak voltage on the MRR, (b) Effect of peak voltage on the EWR, (c) Effect of peak voltage on the SR

Fig. 5.8(b) shows the effect of peak voltage on the EWR. The EWR increases with an increase in peak voltage. This phenomenon can be explained as follows. The single pulse energy, thermal energy density and discharge explosive force increase with an increase in peak voltage, which enhances the material removal of the tool electrode, so the EWR increases.

As shown in Fig. 5.8(c), the SR increases with an increase in peak voltage. The reason for this is that the material removed by a single pulse increases with an increase in peak voltage, the discharge crater becomes larger and deeper; therefore, the SR increases.

5.4.5 Effect of peak current on the process performance

The effect of peak current on the MRR, EWR and SR is shown in Fig. 5.9(a), Fig. 5.9(b) and Fig. 5.9(c), respectively. Fig. 5.9(a) shows the effect of peak current on the MRR. The MRR increases with the increase in peak current. As shown in Fig. 5.9(b), the EWR increases with an increase in peak current. The phenomena can be explained as

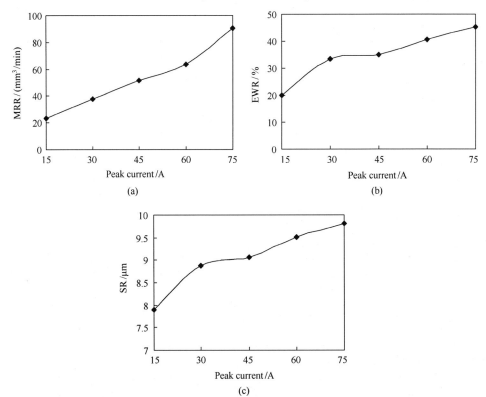

Fig. 5.9 Effect of peak current on the process performance. (a) Effect of peak current on the MRR, (b) Effect of peak current on the EWR, (c) Effect of peak current on the SR

follows: The single pulse energy, thermal energy density and discharge explosive force increase with the increase in peak current, which enhances the material removal of the workpiece and electrode, so the MRR and EWR increase.

The effect of peak current on the SR is shown in Fig. 5.9(c). The SR increases gradually with an increase in peak current. This is because the crater size generated by a single pulse becomes large with an increase in single pulse energy. Single pulse energy increases with the peak current increasing; therefore, the SR increases with an increase in peak current.

5.4.6　Effect of emulsion concentration on the process performance

The effect of emulsion concentration on the MRR, EWR and SR is illustrated in Fig. 5.10. As shown in Fig. 5.10(a), the MRR initially increases with an increase in emulsion concentration, and then decreases with continued increase in emulsion concentration. There are many reasons causing this phenomenon. The dielectric strength, washing capability, density and viscosity of the machining fluid increase with an increase in emulsion concentration, pinch effect and energy density of the discharge channel are enhanced and ejection effect of the eroded material increases; therefore, the MRR rises. However, with a very high viscosity of the machining fluid, the eroded materials are difficult to be flushed away, and the stability of electrical discharges becomes unsatisfactory, so the MRR falls.

Fig. 5.10(b) shows the effect of emulsion concentration on the EWR. The EWR initially decreases with an increase in emulsion concentration, and then increases with continued increase in emulsion concentration. This phenomenon can be explained as follows: There is hydrocarbon in the emulsion, and a deposition layer forms on the electrode surface due to the decomposition of the hydrocarbon during electrical discharge, which can prevent electrode wear. As the emulsion concentration increases, the hydrocarbon content in the dielectric increases and the decomposition of the hydrocarbon and the deposition on the electrode surface are enhanced; therefore, the EWR decreases. However, as the emulsion concentration is higher than 12%, the viscosity of the machining fluid increases largely. The eroded materials are difficult to be flushed away, and they are gathered in the machining zone. The electrical discharge energy supplied to the machining zone repeatedly strikes the un-expelled eroded materials that become concentrated on the machined surface, causing unnecessary electrode wear; therefore, the EWR is high.

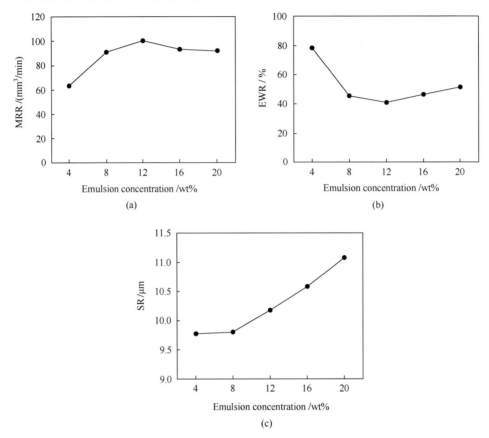

Fig. 5.10 Effect of emulsion concentration on the process performance. (a) Effect of emulsion concentration on the MRR, (b) Effect of emulsion concentration on the EWR, (c) Effect of emulsion concentration on the SR

The effect of emulsion concentration on the SR is shown in Fig. 5.10(c). The SR increases with an increase in emulsion concentration. This is because energy density of the discharge channel increases with an increase in emulsion concentration; the crater size generated by a single pulse becomes larger and deeper; therefore, the SR rises with an increase in emulsion concentration.

5.4.7 Effect of emulsion flux on the process performance

The effect of emulsion flux on the MRR, EWR and SR is illustrated in Fig. 5.11(a)~ Fig. 5.11(c), respectively.

Fig. 5.11(a) shows the influence of emulsion flux on the MRR. The MRR increases with an increase in emulsion flux. This phenomenon can be explained as follows: The high emulsion flux can enhance the cooling effect of the electrode and the ejecting effect of the eroded materials, the electrical discharges become strong and stable; therefore, the MRR increases.

As shown in Fig. 5.11(b), the EWR increases with an increase in emulsion flux. This is because the cooling effect of the electrode and the ejecting effect of the eroded materials are enhanced with an increase in emulsion flux. The deposition effect on the electrode surface decreases and the tool electrode wear increases; therefore, the EWR increases.

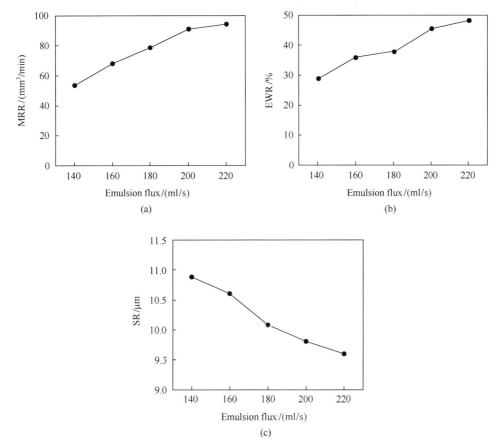

Fig. 5.11 Effect of emulsion flux on the process performance. (a) Effect of emulsion flux on the MRR, (b) Effect of emulsion flux on the EWR, (c) Effect of emulsion flux on the SR

The effect of emulsion flux on SR is shown in Fig. 5.11(c). It can be seen from this figure that the SR decreases with the increase in emulsion flux. This phenomenon can be explained as follows: The high emulsion flux can enhance the cooling effect of the electrode and the ejecting effect of the eroded materials, the eroded materials are flushed away easily, and the discharge points on the workpiece become dispersive and uniform. The craters generated by electrical discharges are shallow and uniformly distributed on the workpiece surface, so the SR decreases with an increase in emulsion flux.

5.4.8 Effect of milling depth on the process performance

The effect of milling depth on the MRR, EWR and SR is illustrated in Fig. 5.12(a)~Fig. 5.12(c), respectively. As shown in Fig. 5.12(a), the MRR initially increases with an increase in milling depth, and then decreases with continued increase in milling depth. There are many reasons causing this phenomenon. Shallower the milling depths with higher discharge current density result in very high thermal energy density of the discharge channel. At this time, the material is removed by vaporization. As the vaporization heat consumption is high, the MRR is low. However, with a very deep milling depth, the discharge current density is low, the SiC ceramic is difficult to be removed by the low current density of

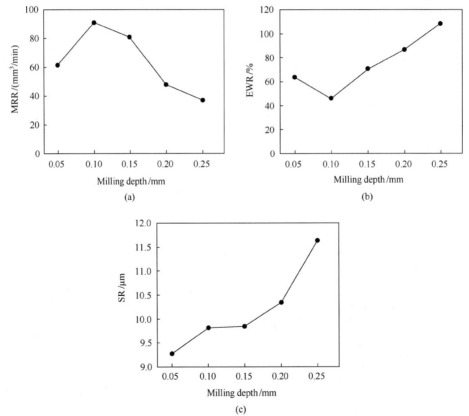

Fig. 5.12 Effect of milling depth on the process performance. (a) Effect of milling depth on the MRR, (b) Effect of milling depth on the EWR, (c) Effect of milling depth on the SR

electrical discharge, and the electrical discharge becomes unstable. Therefore, as the milling depth is greater than 0.1mm, the MRR decreases with an increase in milling depth.

Fig. 5.12(b) shows the effect of milling depth on the EWR. The EWR initially decreases with an increase in milling depth, and then increases with an increase in milling depth. The reason for this is that the shallower the milling depth with higher discharge current density, the thermal energy density and the discharge explosion forces are very high. At this time, the released carbon decomposed from the hydrocarbon in the emulsion is not easily attached to the electrode surface, and the deposition effect on the electrode surface is weak; therefore, the EWR is high. However, with a very deep milling depth, the discharge current density is low, the electrical discharge becomes unstable and arcs are easily generated, causing unnecessary electrode wear; therefore, the EWR is high.

Fig. 5.12(c) shows the influence of milling depth on the SR. The SR increases gradually with an increase in milling depth. The reason for this is that the discharge current density decreases with an increase in milling depth. The electrical discharge becomes unstable and arcs are easily generated; therefore, the SR increases with an increase in milling depth.

5.4.9 Effect of electrode number on the process performance

The effect of electrode number on the MRR, EWR and SR is illustrated in Fig. 5.13(a)~Fig. 5.13(c), respectively. Fig. 5.13(a) shows the effect of electrode number on MRR. The MRR initially increases with an increase in electrode number, and then decreases with continued increase in electrode number. There are many reasons causing this phenomenon. The electrode diameter and the pitch circle diameter in the turntable are constant, as the electrode number is low, the circular pitch between two electrodes is large. The electrical discharge times at a unite time is few, the electrical discharge times at a unite time increases with an increase in electrode number; therefore, the MRR increases with an increase in electrode number. However, when the electrode number is too great, the circular pitch between the two electrodes is very small. The effect of mechanical deionization becomes weak, the electrical discharges become unstable, and arcs are easily produced; therefore, when the electrode number is higher than 6, the MRR decreases as electrode number increases.

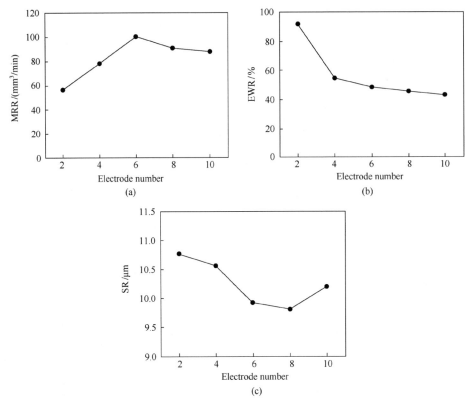

Fig. 5.13 Effect of electrode number on the process performance. (a) Effect of electrode number on the MRR, (b) Effect of electrode number on the EWR, (c) Effect of electrode number on the SR

The effect of electrode number on the EWR is shown in Fig. 5.13(b). The EWR decreases with an increase in electrode number. This phenomenon can be explained as follows: The electrical discharge times at a unite time increases with an increase in electrode number, the discharge energy delivered to the machining gap increases, the dielectric and workpiece are heated for more time, the released carbon decomposed from the hydrocarbon in the emulsion is easily attached to the electrode surface, the deposition effect is enhanced, and the tool electrode wear is compensated, therefore, the EWR decreases.

As shown in Fig. 5.13(c), the SR initially decreases with an increase in electrode number, and then increases with an increase in electrode number. The reason for this is that when the electrode number is little, the circular pitch between two electrodes is large; the cooling effect of the electrode, the effect of mechanical deionization and the ejecting effect of the eroded materials become favorable; electrical discharge explosion force becomes strong; the crater size generated by electric discharge becomes large and deep; therefore, the SR is high. However, with a much greater electrode number, the circular pitch between two electrodes is small. The effect of mechanical deionization becomes weak, the electrical discharges become unstable, and arcs are easily produced; therefore, the SR is high.

5.4.10   Effect of electrode diameter on the process performance

The effect of electrode diameter on the MRR, EWR and SR is illustrated in Fig. 5.14(a)~ Fig. 5.14(c), respectively. Fig. 5.14(a) shows the effect of electrode diameter on the MRR. The MRR initially increases with an increase in electrode diameter and then decreases with an increase in electrode diameter. This is because a smaller electrode diameter with high current density results in very high thermal energy density. At this time, the material is removed by vaporization. As the vaporization heat consumption is high, the MRR is low. However, with a large electrode diameter, the current density and the thermal energy density are low, and electrical discharges become weak; therefore, the MRR is low.

As shown in Fig. 5.14(b), the EWR initially decreases with an increase in electrode diameter and then increases with the continued increase of electrode diameter. The reason for this is that smaller electrode diameter with high discharge current density result in high thermal energy density and discharge explosion force. At this time, the released carbon decomposed from the hydrocarbon in the emulsion does not easily attach to the electrode surface, and the deposition effect on the electrode surface is weak; therefore, the EWR is high. However, with a large electrode diameter the discharge current density is low, the electrical discharge becomes unstable, arcs are easily generated, and cause unnecessary electrode wear; therefore, the EWR is high.

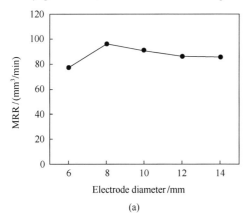

(a)

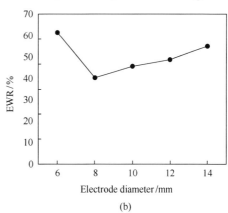

(b)

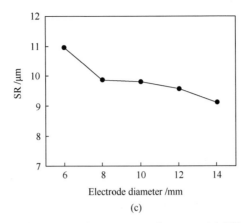

(c)

Fig. 5.14 Effect of electrode diameter on the process performance. (a) Effect of electrode diameter on the MRR, (b) Effect of electrode diameter on the EWR, (c) Effect of electrode diameter on the SR

Fig. 5.14(c) shows the influence of electrode diameter on the SR. The SR decreases with an increase in electrode diameter. This phenomenon can be explained as follows: The current density and the thermal energy density decreases with an increase in electrode diameter, electrical discharges become weak, and the craters generated by electric discharges are little and shallow; therefore, the SR decreases with an increase in electrode diameter.

5.5 Results and discussion of the orthogonal experiment during end ED milling

Table 5.2 presents the design matrix from the L_{25} orthogonal array based on the Taguchi method, as well as the experimental data of the MRR, EWR and SR correlated with each experimental measurement.

Table 5.2 Orthogonal array L_{25} and experimental data

Number of experiment	U/V	I/A	t_i/μs	t_o/μs	MRR /(mm^3/min)	EWR /%	SR /μm
1	90	15	400	300	14.879	9.620	7.011
2	110	30	400	600	36.484	39.907	8.329
3	130	45	400	1200	40.783	45.293	8.729
4	150	60	400	1800	49.784	77.688	8.771
5	170	75	400	2500	54.341	95.775	9.514
6	130	30	120	300	40.611	49.631	8.101
7	150	45	120	600	38.637	60.911	8.214
8	170	60	120	1200	64.600	71.928	8.171

Continued

Number of experiment	U/V	I/A	t_i/μs	t_o/μs	MRR /(mm³/min)	EWR /%	SR /μm
9	90	75	120	1800	25.126	47.375	6.429
10	110	15	120	2500	6.639	40.313	5.014
11	170	45	90	300	70.929	109.919	8.114
12	90	60	90	600	25.200	54.758	6.043
13	110	75	90	1200	25.033	80.916	6.457
14	130	15	90	1800	18.272	38.351	4.514
15	150	30	90	2500	15.687	87.230	6.057
16	110	60	50	300	26.450	58.284	6.143
17	130	75	50	600	52.139	84.690	6.214
18	150	15	50	1200	25.231	67.080	4.814
19	170	30	50	1800	30.164	103.470	6.729
20	90	45	50	2500	12.351	62.292	4.400
21	150	75	40	300	57.253	89.916	6.371
22	170	15	40	600	21.681	77.731	5.614
23	90	30	40	1200	6.404	60.911	4.201
24	110	45	40	1800	15.300	77.852	4.910
25	130	60	40	2500	21.741	97.830	5.529

5.5.1 Analysis of the MRR

The results of ANOVA and F-test values for the MRR are presented in Table 5.3. Open voltage (U) is the significant parameter on the MRR, whereas discharge current (I) is the less significant parameter on the MRR. More electrical energy is conducted into the machining zone within a single discharge upon increasing the open voltage and the discharge current, and more materials could be removed by a single discharge. Therefore, the MRR is high with the high open voltage and the high discharge current.

Table 5.3 ANOVA and F-test of MRR

Parameter	Sum of squares	Degree	Mean square	F
U	3170.179	4	792.5448	7.85**
I	2062.213	4	515.5531	5.11*
t_i	636.037	4	159.0093	1.57
t_o	1119.664	4	279.9161	2.77
Error$_e$	807.731	8	100.9664	
Total	7795.824	24		

* Less significant parameter
** Significant parameter

The average MRR for each factor level is displayed graphically in Fig. 5.15. The average MRR for each factor level indicates the relative effects of the various factors: Open voltage, discharge current, pulse duration (t_i) and pulse interval (t_o). It can be observed from Fig. 5.15 that the MRR increases with the increase in open voltage, discharge current, and pulse duration, respectively, and it decreases with an increase in pulse interval. There are many reasons causing these phenomena. The electrical discharge energy conducted into the machining gap within a single discharge increases with an increase in open voltage, discharge current, and pulse duration, the thermal erosion caused by melting and vaporization are enhanced on the machined surface. The material removal is intensified, therefore, the MRR increases. In addition, the electrical discharge frequency decreases with an increase in pulse interval, and less materials could be removed in a unit time, so the MRR decreases with an increase in pulse interval. The optimal combination levels of machining parameters that maximized MRR of end ED milling on SiC ceramic from Fig. 5.15 are as follows: 170V open voltage, 75A discharge current, 400μs pulse duration, and 300μs pulse interval.

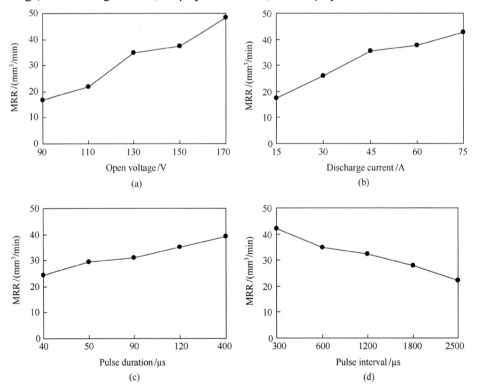

Fig. 5.15 Effects of machining parameters on the average MRR. (a) Effect of open voltage on the average MRR, (b) Effect of discharge current on the average MRR, (c) Effect of pulse duration on the average MRR, (d) Effect of pulse interval on the average MRR

The capability of the stepwise model to represent the experimental data is assessed through the analysis of variance. The results of the analysis of variance for the stepwise model of MRR are shown in Table 5.4. As in this case F is far greater than $F_{0.001,5,19}$, it is accepted that the mathematical model of the MRR by using stepwise regression analysis method is appropriate for a significance level. The equation for the mathematical model is as follows,

$$\text{MRR}=18.87338+0.00097728 t_i\, U - 0.00000180 t_i^2 U - 2.51237 \times 10^{-8} t_o^2 U \\ - 0.00001978 I^2 U + 0.00003605 I U^2 \tag{5.1}$$

Table 5.4 ANOVA table for the stepwise regression model of the MRR

Source	DF	Sum of squares	Mean square	F value	$F_{0.001,5,19}$
Model	5	7047.6149	1409.5230	35.19	6.62
Error	19	748.2098	39.3795		
Corrected Total	24	7795.8247			

Fig. 5.16 shows the estimated response surface for the MRR parameter in relation to the design parameters of open voltage and discharge current for the pulse duration of 400μs, and pulse interval of 300μs. The MRR parameter tends to increase with an increase in the open voltage for any value of the discharge current factor, and increase with the increase of the discharge current for any value of the open voltage. Furthermore, the maximal MRR can be obtained with the maximal open voltage and the maximal discharge current.

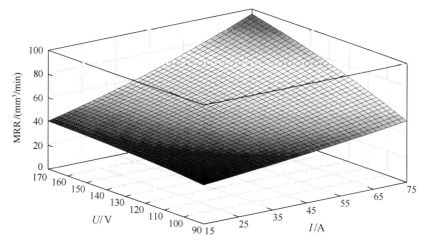

Fig. 5.16 Estimated response surface of the MRR versus U and I

5.5.2 Analysis of the EWR

The results of ANOVA and F-test values for the EWR are presented in Table 5.5. Open voltage and pulse duration are the significant parameters on the EWR, whereas discharge current is the less significant parameter on the EWR.

Table 5.5 ANOVA and F-test of the EWR

Parameter	Sum of squares	Degree	Mean square	F
U	5883.969	4	1470.992	12.80**
I	3106.099	4	776.5248	6.76*
t_i	3280.544	4	820.1361	7.14**
t_o	615.190	4	153.7976	1.34
Error$_e$	919.138	8	114.8923	
Total	13804.940	24		

* Less significant parameter
** Significant parameter

Fig. 5.17 shows the average EWR for each factor level. The average EWR for each factor level indicates the relative effects of the various factors: Open voltage, discharge current, pulse duration, and pulse interval. The EWR increases with an increase in open voltage, discharge current, and pulse interval, respectively, and it decreases with an increase in pulse duration. There are many reasons causing the phenomena. The single pulse energy, thermal energy density and discharge explosive force increase with an increase in open voltage and discharge current, which enhances the material removal of the tool electrode, so the EWR increases with an increase in open voltage, and discharge current, respectively. During EDM, a deposition layer can form on the electrode surface due to the decomposition of the dielectric and workpiece material attached to the tool electrode surface, and the tool electrode wear can be prevented by the protective effects of the deposition layer [28,30]. As the pulse duration increases, the discharge energy delivered to the machining gap increases, the dielectric and workpiece are heated for more time, the released carbon decomposed from the dielectric is easily attached to the electrode surface, the deposition effect is enhanced, the tool electrode wear decreases; therefore, the EWR decreases. As the pulse interval increases, the time for deionization of the dielectric increases, the discharge energy delivered to the machining gap decreases in a unit time, the released carbon decomposed from the dielectric decreases, the deposition effect weakens, the tool wear compensation decreases; therefore, the EWR increases. The optimal combination levels of machining parameters that minimized the EWR of end ED milling on SiC ceramic from Fig. 5.17 are as follows: 90V open voltage, 15A discharge current, 400μs pulse duration, and 300μs pulse interval.

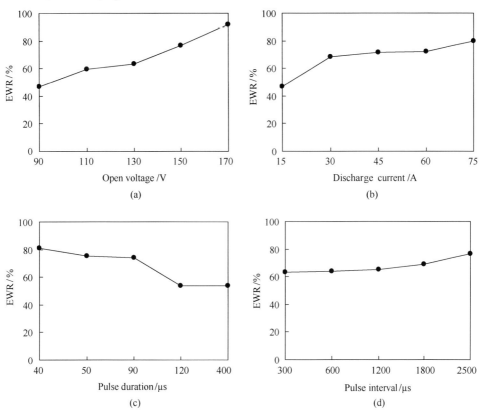

Fig. 5.17 Effects of machining parameters on the average EWR. (a) Effect of open voltage on the average EWR, (b) Effect of discharge current on the average EWR, (c) Effect of pulse duration on the average EWR, (d) Effect of pulse interval on the average EWR

The results of the analysis of variance for the stepwise model of EWR are shown in Table 5.6. As in this case F is far greater than $F_{0.001,7,17}$, it is accepted that the mathematical model of the EWR by using stepwise regression analysis method is appropriate for a significance level. The mathematical relation between the EWR and machining parameters is as follows.

$$EWR = 35.84721 - 0.12215 t_i - 0.00001646 t_i^2 U + 0.00014255 I U^2 \\ + 4.266302 \times 10^{-8} t_i^3 U + 6.61759 \times 10^{-12} t_o^3 U + 0.00000132 I^3 U \quad (5.2) \\ - 0.00000205 I^2 U^2$$

Table 5.6 ANOVA table for the stepwise regression model of the EWR

Source	DF	Sum of Squares	Mean Square	F Value	$F_{0.001,7,17}$
Model	7	12645.0748	1806.4308	26.48	6.22
Error	17	1159.9252	68.2309		
Corrected Total	24	13805			

The response surface of the EWR parameter as a function of the open voltage and pulse duration for a discharge current of 15A and pulse interval of 300μs, is shown in Fig. 5.18. The figure indicates that the value of the EWR parameter tends to increase with an increase in the open voltage for any value of the pulse duration factor, and decreases with an increase in the pulse duration for any value of the open voltage. Furthermore, the minimal EWR can be obtained with the minimal open voltage and the maximal pulse duration.

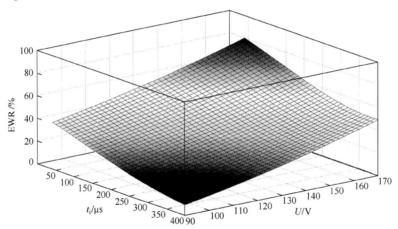

Fig. 5.18 Estimated response surface of EWR versus U and t_i

5.5.3 Analysis of the SR

The results of the ANOVA and F-test values for the SR are presented in Table 5.7. Pulse duration, open voltage and discharge current are the more significant parameters on the SR, whereas pulse interval is the significant parameter on the SR. When the pulse duration, open voltage, and discharge current are set at the high level, the huge discharge energy would be delivered into the machining zone within a single pulse, and the discharge crater becomes larger and deeper. Therefore, the SR is high using high pulse duration, open voltage, and discharge current, respectively.

Table 5.7 ANOVA and F-test of the SR

Parameter	Sum of squares	Degree	Mean square	F
U	11.366	4	2.8415	31.86**
I	9.028	4	2.2570	25.30**
t_i	32.389	4	8.0973	90.78**
t_o	3.737	4	0.9343	10.48*
Error$_e$	0.714	8	0.0892	
Total	57.234	24		

* Significant parameter
** More significant parameter

The average SR for each factor level is displayed graphically in Fig. 5.19. The average SR for each factor level indicates the relative effects of the various factors: Open voltage, discharge current, pulse duration and pulse interval. It can be observed from Fig. 5.19 that the SR increases with an increase in open voltage, discharge current, and pulse duration, respectively, and it decreases with an increase in pulse interval. The phenomenon can be explained as follows: The material removed by a single pulse increases with an increase in open voltage, discharge current, and pulse duration, respectively, the discharge crater becomes larger and deeper; therefore, the SR increases. In addition, with a longer pulse interval, there is more time to clear the disintegrated particles from the discharge gap, and enhance deionization function of the dielectric; therefore, the SR is low. The optimal combination levels of machining parameters that minimized the SR of end ED milling on SiC ceramic from Fig. 5.19 are as follows: 90V open voltage, 15A discharge current, 40μs pulse duration, and 2500μs pulse interval.

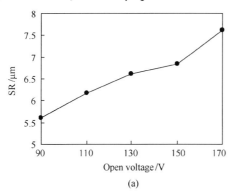

(a)

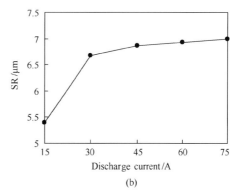

(b)

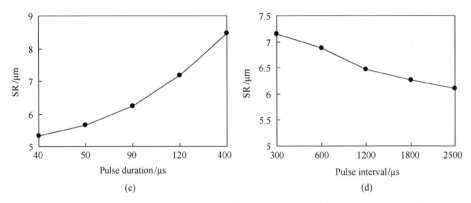

Fig. 5.19 Effects of machining parameters on the average SR. (a) Effect of open voltage on the average SR, (b) Effect of discharge current on the average SR, (c) Effect of pulse duration on the average SR, (d) Effect of pulse interval on the average SR

The results of the analysis of variance for the stepwise model of SR are shown in Table 5.8. As in this case F is far greater than $F_{0.001,6,18}$, it is accepted that the mathematical model of SR by using stepwise regression analysis method is appropriate for a significance level. The equation for the mathematical model is as follows:

$$SR = 3.84691 - 0.00109 t_o + 0.04544 I - 2.63404 \times 10^{-8} t_i^2 t_o + 2.306377 \times 10^{-9} t_o^2 U \\ + 0.00000134 t_i U^2 - 1.53678 \times 10^{-8} I^2 U^2 \qquad (5.3)$$

Table 5.8 ANOVA table for the stepwise regression model of the SR

Source	DF	Sum of squares	Mean square	F value	$F_{0.001,6,18}$
Model	6	53.8852	8.9809	48.27	6.35
Error	18	3.3488	0.1861		
Corrected Total	24	57.2340			

Fig. 5.20 shows the estimated response surface for the SR parameter in relation to the design parameters of open voltage and pulse duration for the discharge current of 15A, and pulse interval of 2500μs. As can be seen in this figure, the SR parameter tends to increase with an increase in the open voltage for any value of the pulse duration factor, and increases with an increase in the pulse duration for any value of the open voltage. Furthermore, the minimal SR can be obtained with the minimal open voltage and the minimal pulse duration.

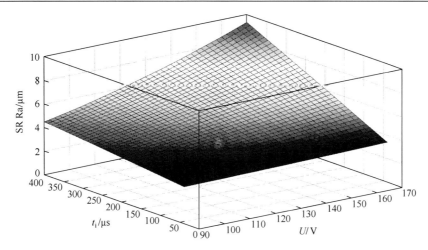

Fig. 5.20 Estimated response surface of SR versus U and t_i

5.5.4 Confirmation experiments

The purpose of confirmation experiment is to validate the conclusions drawn in the previous analysis. The confirmation experiments for the MRR, EWR and SR are conducted at the optimum setting of the machining parameters, and the results are shown in Table 5.9. As shown in this table, the experimental values agree well with predictions, and they are improved significantly in comparison with the experimental values from Table 5.2, which confirms the validity of the used Taguchi method for enhancing the machining performance and optimizing the machining parameters.

Table 5.9 Results of the confirmation experiment for end ED milling of SiC ceramic

	Optimal machining parameters				Predicted	Experimental	Relative error/%
	U/V	I/A	$t_i/\mu s$	$t_o/\mu s$			
MRR/(mm³/min)	170	75	400	300	95.208	94.743	0.49
EWR/%	90	15	400	300	9.703	9.620	0.86
SR/μm	90	15	40	2500	3.402	3.453	1.48

5.6 Analysis of the machined surface by end ED milling

5.6.1 SEM observation of the machined surface

The SEM micrographs of the machined surface with different tool polarities are illustrated in Fig. 5.21. Many spark-induced craters, melt-formation droplets and micropores exist in these micrographs. The phenomena indicate that the SiC ceramic is molten or evaporated by the sparking thermal energy. During machining, sparks are formed at the electrical conductive phase on the SiC ceramic. The discharged energy produces very high temperatures at the point of the spark, causing a minute part of the workpiece to melt and vaporize. As the spark ceases, a violent bubble causes superheated, molten liquid on the workpiece surface to explode into the gap. Most molten material is flushed away to form craters on the surface. However, not all of the molten material can be removed because of the surface tension, tensile strength, and bonding forces between liquid and solid. The molten material remained on the surface of workpiece is cooled by the working fluid, and some droplets are formed. The formation of micropores is ascribed to the ejection of gases that escape from the solidified material.

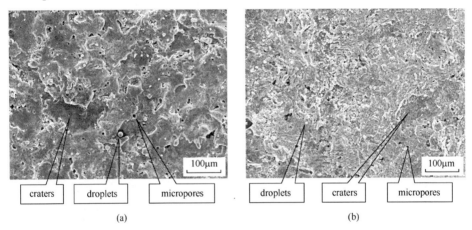

Fig. 5.21 SEM micrographs of SiC ceramic surfaces machined by end ED milling with different tool polarities. (a) tool(−), (b) tool(+)

It is obvious that the craters are bigger and deeper in case of negative tool polarity, when compared to positive tool polarity under the same conditions. As mentioned above, the bombardment effect by the electrons is stronger than that by positive ions during end ED milling SiC ceramic. Stronger electrons bombardment results in bigger and deeper craters, leading to a rougher surface.

The machined surfaces of SiC ceramic sample in different machining conditions are observed under SEM with higher magnification. Fig. 5.22(a) and Fig. 5.22(b) show the typical micrographs obtained from the observation. It is evident that the surface micro-cracks are observed on the machined surface. The micro-cracks formation is associated with the development of high thermal stresses during machining, and it is responsible for the thermal spalling removal of the SiC ceramic.

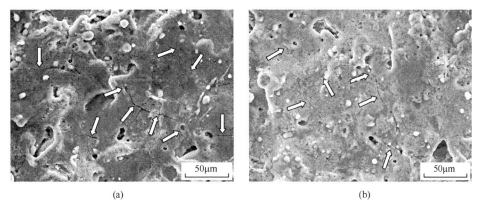

Fig. 5.22 SEM micrograph showing micro-cracks on the machined surface with negative tool polarity in different machining conditions. The arrows indicate micro-cracks. (a) medium, (b) fine

In the end ED milling process, the SiC ceramic goes through thermal cycling, so a complex temperature gradient is established. On sudden heating, local compressive shear stresses develop because the expanding material is prevented from doing so by the cooler interior material. At the same time, this places the interior material in tension as it is pulled by the outer material as it tries to expand. This situation is reversed on sudden cooling. The SiC ceramic workpiece is subjected to higher stress due to more severe gradients. When the degree of induced stress exceeds the maximum tensile strength of the SiC ceramic, cracking occurs on the machined surface. With extreme hardness and brittleness, the SiC ceramic tends to promote the formation of steep temperature gradients away from the melting and evaporation zone, so the micro-cracks are easily produced, which leads to thermal spalling of the SiC ceramic during machining. The removal mechanism in end ED milling of SiC ceramic consists of not just the melting and evaporation but also thermal spalling.

It can also be seen from Fig. 5.22 that the number and length of the micro-cracks on the machined surface increase from fine machining to medium machining. The discharge energy delivered to the machining gap increases from fine machining to medium machining, the heating and cooling effect is enhanced, the temperature gradient and stress during machining increase, so the number and length of the micro-cracks on the surface increase from fine machining to medium machining.

5.6.2 Compositions of the machined surface

During electric discharge machining, the material can be transferred between the electrodes in solid, molten or gaseous state simultaneously [31,32]. EDS spectrum analysis is used to identify the elements on the workpiece surface in different machining conditions.

As shown in Fig. 5.23, Cu is present on the machined surface, whereas it is not detected

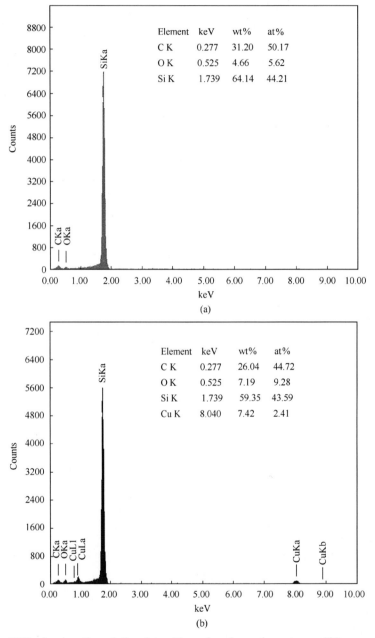

Fig. 5.23 EDS showing the relative intensities of various elements on SiC ceramic surface. (a) Unprocessed SiC ceramic surface, (b) SiC ceramic surface created at coarse machining mode with positive tool polarity

on the unprocessed surface. This can be explained by the melting and resolidification of the copper electrode during end ED milling spark erosion. It implies that some tool electrode material transfers to the workpiece surface during machining.

The copper percentage on machined SiC ceramic surface with different machining conditions is shown in Fig. 5.24. The copper percentage on the machined surface increases from fine machining to coarse machining. This is because the single pulse energy, thermal energy density and discharge explosive force increase from fine machining to coarse machining, the electrode material removal is enhanced, and more electrode material can transfer to the workpiece surface; therefore, the copper percentage on the machined surface increases from fine machining to coarse machining. It can also be seen from Fig. 5.24 that under the same conditions the copper percentage on the machined surface in positive tool polarity is 1.2~4.6 times higher than that in negative tool polarity. This phenomenon can be explained as follows: The copper is a metal, the electrolysis reaction occurs easily under the application of the water-based emulsion in positive tool polarity, and the removed copper electrode is mostly ionized into Cu^{2+}. The Cu^{2+} can attach to the workpiece surface easily, which is connected to the negative pole of the pulse generator. However, the electrolysis reaction occurs difficultly in negative tool polarity because the SiC is a nonmetal. Therefore, the copper percentage on the machined surface in positive tool polarity is higher than that in negative tool polarity.

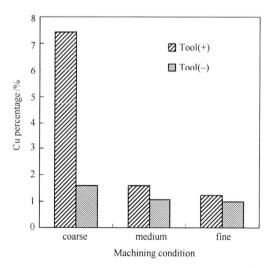

Fig. 5.24 Copper percentage on SiC ceramic surface machined by end ED milling with different machining conditions

XRD is used to further analyze the material composition of the workpiece surface.

The XRD results are shown in Fig. 5.25. Major peaks in the scan reveal that the workpiece surface at coarse machining mode with positive tool polarity contains SiC, Si and copper silicide (Cu_3Si). Cu_3Si is not observed on the unprocessed surface. This can be explained that a modified layer forms on the machined surface, and a combination reaction takes place during end ED milling of SiC ceramic, which can be described as follows.

$$3Cu+Si=Cu_3Si \tag{5.4}$$

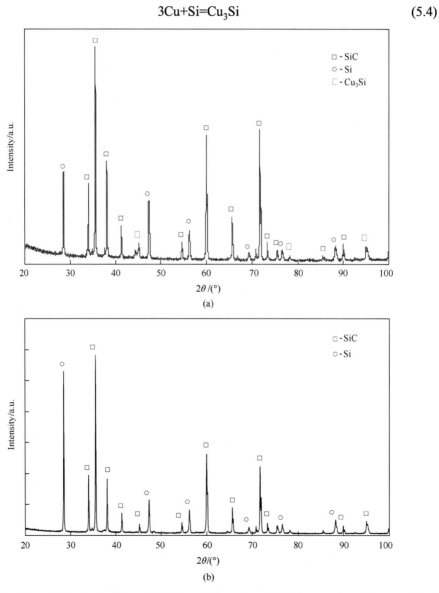

Fig. 5.25 X-ray diffraction patterns of SiC ceramic surface. (a) SiC ceramic surface created at coarse machining mode with positive tool polarity, (b) Unprocessed SiC ceramic surface

5.7 Conclusions

Using a large turntable with several small cylindrical copper rods as the tool electrode, the weakly conductive SiC ceramic with the electrical resistivity of 500Ω·cm can be easily machined by end ED milling. The process shows high MRR. In addition, end ED milling employs a water-based emulsion as the machining fluid, and it shows a good working environmental practice.

With the negative tool polarity and the longer pulse duration, the higher MRR, lower EWR and higher SR can be obtained. With the higher peak voltage and the higher peak current, the higher MRR, higher EWR and higher SR can be obtained. During machining, the shorter pulse interval should be used.

With a suitable emulsion concentration, higher MRR and lower EWR can be obtained. However, the SR increases with an increase in emulsion concentration. The MRR and the EWR increase with an increase in emulsion flux, whereas the SR decreases with an increase in emulsion flux. During end ED milling of SiC ceramic, higher MRR results in lower EWR, and lower SR can be acquired with the appropriate milling depth, electrode diameter and electrode number.

ANOVA and F-test of experimental data related to the essential machining parameters of end ED milling reveal that open voltage is the significant parameter on the MRR, whereas discharge current is the less significant parameter on the MRR. Open voltage and pulse duration are the significant parameters on the EWR, whereas discharge current is the less significant parameter on the EWR. Pulse duration, open voltage and discharge current are the more significant parameters on the SR, whereas pulse interval is the significant parameter on the SR.

The analysis of experimental data based on the Taguchi method reveals that the optimal combination levels of machining parameters that maximized the MRR are as follows: 170V open voltage, 75A discharge current, 400μs pulse duration, and 300μs pulse interval.

The analysis of experimental data based on the Taguchi method reveals that the optimal combination levels of machining parameters that minimized the EWR are as follows: 90V open voltage, 15A discharge current, 400μs pulse duration, and 300μs pulse interval.

The analysis of experimental data based on the Taguchi method reveals that the optimal combination levels of machining parameters that minimized the SR are as follows: 90V open voltage, 15A discharge current, 40μs pulse duration, and 2500μs pulse interval.

With each optimal combination level of machining parameters, the MRR, EWR and SR can reach 94.743mm^3/min, 9.620% and 3.453μm, respectively, which agree well with predictions. The experimental results confirm the validity of the used Taguchi method for enhancing the machining performance and optimizing the machining parameters.

The machined surfaces are characterized by craters, droplets, micropores and micro-cracks. The removal mechanism during end ED milling of SiC ceramic consists of melting, evaporation and thermal spalling. The material from the tool electrode can transfer to the workpiece surface, and a combination reaction takes place during machining. Furthermore, the tool electrode material that transfers to the workpiece surface increases from fine machining to coarse machining, and more tool electrode material can transfer to the workpiece surface in case of positive tool polarity, when compared to negative tool polarity under the same conditions.

References

[1] Guo X Z, Yang H, Zhang L J, et al. Sintering behavior, microstructure and mechanical properties of silicon carbide ceramics containing different nano-TiN additive. Ceramics International, 2010, 36(1): 161-165.

[2] Bae H T, Choi H J, Jeong J H, et al. The effect of reaction temperature on the tribological behavior of the surface modified silicon carbide by the carbide derived carbon process. Materials and Manufacturing Processes, 2010, 25(5): 345-349.

[3] Yu X M, Zhou W C, Luo F, et al. Effect of fabrication atmosphere on dielectric properties of SiC/SiC composites. Journal of Alloys and Compounds, 2009, 479(1/2): L1-L3.

[4] Okada A. Automotive and industrial applications of structural ceramics in Japan. Journal of the European Ceramic Society, 2008, 28(5): 1097-1104.

[5] Krnel K, Stadler Z, Kosmač T. Carbon/carbon–silicon-carbide dual-matrix composites for brake discs. Materials and Manufacturing Processes, 2008, 23(6): 587-590.

[6] Katoh Y, Kondo S, Snead L L. DC electrical conductivity of silicon carbide ceramics and composites for flow channel insert applications. Journal of Nuclear Materials, 2009, 386-388(C): 639-642.

[7] Zhong Z W. Grinding of toroidal and cylindrical surfaces on SiC using diamond grinding wheels. Materials and Manufacturing Processes, 1997, 12(6): 1049-1062.

[8] Zhong Z W, Ramesh K, Yeo S H. Grinding of nickel-based super-alloys and advanced ceramics. Materials and Manufacturing Processes, 2001, 16(2): 195-207.

[9] Agarwal S, Rao P V. Experimental investigation of surface/subsurface damage formation and material removal mechanisms in SiC grinding. International Journal of Machine Tools and

Manufacture, 2008, 48 (6): 698-710.

[10] Yin L, Vancoille E Y J, Lee L C, et al. High-quality grinding of polycrystalline silicon carbide spherical surfaces. Wear, 2004, 256 (1/2): 197-207.

[11] Gopal A V, Rao P V. A new chip-thickness model for performance assessment of silicon carbide grinding. International Journal of Advanced Manufacturing Technology, 2004, 24 (11/12): 816-820.

[12] Zhang Z, Yan J, Kuriyagawa T. Machinability investigation of reaction-bonded silicon carbide by single-point diamond turning. Key Engineering Materials, 2009, 389/390: 151-156.

[13] Yamaguchi S, Noro T, Takahashi H, et al. Electric discharge machining for silicon carbide and related materials. Materials Science Forum, 2009, 600-603: 851-854.

[14] Seo Y W, Kim D, Ramulu M. Electrical discharge machining of functionally graded 15-35vol% SiC_p/Al composites. Materials and Manufacturing Processes, 2006, 21(5): 479-487.

[15] Mahdavinejad R, Tolouei R M, Sharifi B H. Heat transfer analysis of EDM process on silicon carbide. International Journal of Numerical Methods for Heat and Fluid Flow, 2005, 15(5): 483-502.

[16] Kato T, Noro T, Takahashi H, et al. Characterization of electric discharge machining for silicon carbide single crystal. Materials Science Forum, 2009, 600-603: 855-858.

[17] Luis C J, Puertas I, Villa G. Material removal rate and electrode wear study on the EDM of silicon carbide. Journal of Materials Processing Technology, 2005, 164/165: 889-896.

[18] Shih H R, Shu K M. A study of electrical discharge grinding using a rotary disk electrode. International Journal of Advanced Manufacturing Technology, 2008, 38(1/2): 59-67.

[19] Lauwers B, Kruth J P, Brans K. Development of technology and strategies for the machining of ceramic components by sinking and milling EDM. CIRP Annals-Manufacturing Technology, 2007, 56(1): 225-228.

[20] Chang Y F, Hong R C. Parametric curve machining of a CNC milling EDM. International Journal of Machine Tools and Manufacture, 2005, 45(7/8): 941-948.

[21] Han F Z, Wang Y X, Zhou M. High-speed EDM milling with moving electric arcs. International Journal of Machine Tools and Manufacture, 2009, 49(1): 20-24.

[22] Liu Y H, Ji R J, Li Q Y, et al. An experimental investigation for electric discharge milling of SiC ceramics with high electrical resistivity. Journal of Alloys and Compounds, 2009, 472(1/2): 406-410.

[23] Kunieda M, Miyoshi Y, Takaya T, et al. High speed 3D milling by dry EDM. CIRP Annals - Manufacturing Technology, 2003, 52(1): 147-150.

[24] Kao J Y, Tsao C C, Wang S S, et al. Optimization of the EDM parameters on machining Ti-6Al-4V with multiple quality characteristics. International Journal of Advanced Manufacturing

Technology, 2010, 47(1/4): 395-402.

[25] Ponappa K, Aravindan S, Rao P V, et al. The effect of process parameters on machining of magnesium nano alumina composites through EDM. International Journal of Advanced Manufacturing Technology, 2010, 46(9/12): 1035-1042.

[26] Pal S, Malviya S K, Pal S K, et al. Optimization of quality characteristics parameters in a pulsed metal inert gas welding process using grey-based Taguchi method. International Journal of Advanced Manufacturing Technology, 2009, 44(11/12): 1250-1260.

[27] Wang Q H, Chang Z G, Li R H. Random data processing method. Dongying: China University of Petroleum Press, 2005.

[28] Marafona J. Black layer characterisation and electrode wear ratio in electrical discharge machining (EDM). Journal of Materials Processing Technology, 2007, 184(1/3): 27-31.

[29] Kunieda M, Kobayashi T. Clarifying mechanism of determining tool electrode wear ratio in EDM using spectroscopic measurement of vapor density. Journal of Materials Processing Technology, 2004, 149(1/3): 284-288.

[30] Marafona J D. Black layer affects the thermal conductivity of the surface of copper-tungsten electrode. International Journal of Advanced Manufacturing Technology, 2009, 42(5/6): 482-488.

[31] Guu Y H, Hou M T K. Effect of machining parameters on surface textures in EDM of Fe-Mn-Al alloy. Materials Science and Engineering A, 2007, 466:61-67.

[32] Chen Y F, Lin Y C. Surface modifications of Al-Zn-Mg alloy using combined EDM with ultrasonic machining and addition of TiC particles into the dielectric. Journal of Materials Processing Technology, 2009, 209: 4343-4350.

Chapter 6   Electric Discharge Milling and Mechanical Grinding Compound Machining of Weakly Conductive Engineering Ceramics

Weakly conductive engineering ceramics, such as SiC ceramics, have been widely used in modern industry. However, the manufacture of SiC ceramics is not an efficient process. This chapter proposes a new technology of machining SiC ceramics with electrical discharge milling and mechanical grinding compound method. The compound process employs the pulse generator used in electrical discharge machining, and uses a water-based emulsion as the machining fluid. It is able to effectively machine a large surface area on SiC ceramics with a good surface quality. In this chapter, the effects of pulse duration, pulse interval, peak voltage, peak current and feed rate of the workpiece on the process performance such as the MRR, the relative electrode wear ratio (REWR) and the SR have been investigated. A L_{25} orthogonal array based on Taguchi method is adopted, and the experimental data are statistically evaluated by analysis of variance and stepwise regression. The significant machining parameters, the optimal combination levels of machining parameters, and the mathematical models associated with the process performance are obtained. In addition, the surface microstructures machined by the new process have been observed by SEM, XRD and EDS.

6.1   Introduction

Weakly conductive engineering ceramics, such as SiC ceramics, have the advantages of light weight, chemical and thermal stabilities at elevated temperatures and excellent wear resistance, in comparison with steel [1]. Because of their advantages, SiC ceramics have been widely used in optical mirror, accelerometer, refractories, electronic components, and in the biomedical, aerospace, and defense industries [2-4].

However, SiC ceramics are the typically difficult-to-machine materials owing to the beneficial properties such as high intensity, high rigidity, superior wear resistance, etc. Diamond grinding and diamond turning are the most commonly used techniques to machine SiC ceramics with a good surface quality, but they are costly and inefficient. Moreover, the high hardness of SiC ceramics induces higher cutting force and quick wear of diamond cutting edges [5].

Some other reported studies on machining SiC ceramics include electrical discharge machining [6-8], electrical discharge milling [9-11], machining with diamond tools [12-14], ultrasonic machining [15,16], plasma chemical vaporization machining [17-18], laser beam machining etc [19,20]. However, the literature review states that difficulty, high cost and long time associated with machining SiC ceramics limit the use of SiC in industry.

This chapter proposes a new technique of machining SiC ceramics with electrical discharge milling and mechanical grinding compound method. The process employs the pulse generator used in EDM, and uses a water-based emulsion as the machining fluid. Electrical discharge milling and mechanical grinding happen alternately and they are mutually beneficial, so it is able to effectively machine a large surface area on SiC ceramics with a good surface quality. A L_{25} orthogonal array based on Taguchi method is adopted. The effects of pulse duration, pulse interval, peak voltage, peak current and feed rate of the workpiece on the process performance have been investigated. The experimental data are statistically evaluated by the ANOVA and stepwise regression.

6.2 Principle for ED milling and mechanical grinding of SiC ceramic

The principle for ED milling and mechanical grinding of SiC ceramic is shown in Fig. 6.1. The tool and the workpiece are connected to the positive and negative poles of the pulse generator, respectively. The tool is a steel wheel with uniform-distributed abrasive sticks in the circumference and is mounted on to a rotary spindle, driven by an A.C. motor. The workpiece is SiC ceramic blank and is mounted on to a NC table. The machining fluid is a water-based emulsion.

During machining, the tool rotates at a high speed; the machining fluid is flushed into the gap between the tool and the workpiece with a nozzle, the SiC ceramic workpiece is fed towards the tool with NC table. As the workpiece approaches the tool and the distance between the workpiece and the steel tooth reaches the discharge gap, electrical discharges are produced. A high temperature and pressure plasma channel grows rapidly during the pulse duration [21]. The instantaneous high temperature and

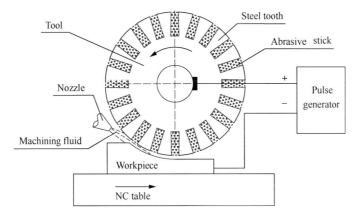

Fig. 6.1 Schematic illustration for ED milling and mechanical grinding of SiC ceramic

pressure plasma cause SiC ceramic to be removed by electrical discharge milling, and a modified surface layer is formed on the workpiece. The following abrasive stick grinds the modified surface layer. The modified surface layer can be removed easily by abrasive stick. Electrical discharge milling and mechanical grinding happen alternately and they are mutually beneficial, so the high material removal rate and good surface quality of machining SiC ceramic can be achieved. Furthermore, the tool is manufactured easily and it shows low cost.

The advantage of the compound process over conventional electric discharge machining and mechanical grinding can be explained as follows. During conventional electric discharge machining of SiC ceramic, the modified surface layer forms easily, which makes the subsequent discharge difficult and debases the workpiece surface quality. During conventional mechanical grinding of SiC ceramic, the workpiece is hard and the grinding force is high, so the material removal rate is low. During this compound machining of SiC ceramic, electrical discharge milling and mechanical grinding happen alternately and they are mutually beneficial, so the high material removal rate and good surface quality can be achieved simultaneously. The comparison experiments have been done, for positive tool polarity, pulse duration of 50μs, pulse interval of 1000μs, peak current of 25A, peak voltage of 200V, and feed rate of the workpiece of 20cm/s. The experimental results are shown in Fig. 6.2. The MRR, the lower SR and the lower REWR can be obtained with the compound process in comparison with conventional EDM. Furthermore, although the SR with the compound process is a little higher than that with conventional diamond grinding, the MRR is far higher and the REWR is far lower with the compound process in comparison with conventional diamond grinding.

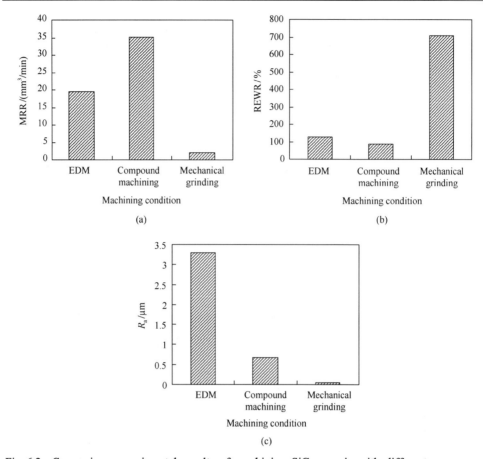

Fig. 6.2 Comparison experimental results of machining SiC ceramic with different processes. (a) Effect of machining condition on the MRR, (b) Effect of machining condition on the REWR, (c) Effect of machining condition on the SR

6.3 Experiments

6.3.1 Experimental procedures

The experiments are carried out on an electrical discharge grinding machine. In the following experiments, the workpiece material is the SiC ceramic with the dimension of 50mm×50mm×8mm, and the machining fluid is a water-based emulsion composed of 5wt% emulsified oil and 95wt% distilled water. The tool is a steel wheel with twenty resin bonded abrasive sticks in the circumference, the ratio of sticks thickness and circular pitch for the tool is 1∶2, the diameter of the wheel is 210mm, and the width of the wheel is 15mm. According to the former experiments, the machining performance is improved with a high wheel speed. The maximum spindle speed of the electrical discharge grinding machine is 3000r/min, so the wheel speed is set 3000r/min. The tool photograph is shown in Fig. 6.3.

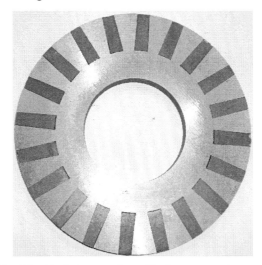

Fig. 6.3 Tool photograph used in the compound process

The tool polarity is negative. The SR is measured by a surface roughness tester. The microstructure of the workpiece surface is observed by SEM, and equipped with EDS. The workpiece removal volume and electrode wear volume are obtained through measuring the dimension of the workpiece and the electrode before and after machining with two dial indicators and a vernier caliper. The least count of the dial indicator is 0.001mm, and the least count of the vernier caliper is 0.01mm. The dial indicator named m_1 is fixed on a magnetic stand. The dial indicator named m_2 is fixed on the electrical discharge grinding machine, and it moves along with the wheel in the vertical direction. At first, an initial groove is machined on the workpiece. The initial depth of the groove is measured with m_1, and the scale value of the m_1 (h_0) is obtained to represent the depth of the groove before machining. At this time, the scale value of m_2 (H_0) is also obtained to represent the initial location of the wheel. After machining, the depth of the groove is measured with m_1, and the scale value of m_1 (h_1) is obtained to represent the depth of the groove after machining. At this time, the scale value of m_2 (H_1) is also obtained to represent the location of the wheel after machining. The absolute value of h_1 minus h_0 name Δh was the depth change of the groove during machining, the length and the width of the groove are measured by a vernier caliper, and the material removal volume during machining can be obtained, therefore, the MRR could be calculated. If the steel in the wheel wear more, the abrasive will be protruded, at this time the grinding function enhance, which can make the wheel homogeneous. Similarly, if the abrasive in the wheel wear more, the steel will be protruded, at this time the ED milling enhance, which can also make the wheel homogeneous. The absolute value of H_1 minus H_0 name ΔH is the fall of the wheel spindle, and the reduction of the wheel radius name ΔR equaled to ΔH minus Δh. ΔR is far less than the initial wheel diameter, the wheel wear volume (WWV) during machining can be expressed as follow:

$$\text{WWV} = \pi d \Delta R w \qquad (6.1)$$

Where, d is the initial wheel diameter, and w is the wheel width.

The SR is measured by a surface roughness tester. The workpiece surface morphology is examined with SEM, XRD and EDS. All the observed specimens have been cleaned ultrasonically and dried with a hot-air blower before the examination.

6.3.2 Experimental design

In this investigation, the experimental design is according to a L_{25} orthogonal array based on the Taguchi method. Five machining parameters such as pulse duration, pulse interval, peak voltage, peak current and feed rate of the workpiece (V_f) were varied to determine their effects on the machining characteristics in terms of the MRR, REWR and SR. Five levels for the above-mentioned every parameter are selected. The machining parameters and their levels used in this investigation are given in Table 6.1. There is no confounding between the main effects, so the resolution in this Taguchi design is III. The experimental data are statistically evaluated by the ANOVA, and a series of mathematical models are obtained by using the stepwise regression method with a commercially available mathematical software SAS (version 8.0).

Table 6.1 Machining parameters and their levels

Parameters	Levels				
	1	2	3	4	5
$t_i/\mu s$	500	200	100	50	40
$t_o/\mu s$	500	1000	1500	2000	2500
I/A	5	10	15	20	25
U/V	125	150	175	200	225
$V_f/(cm/s)$	20	17.5	15	12.5	10

6.4 Experimental results and discussion of orthogonal array

The design matrix from the L_{25} orthogonal array based on the Taguchi method, and the experimental data of the MRR, REWR and SR are shown in Table 6.2.

Table 6.2 Orthogonal array L_{25} and experimental data

Number of experiment	$t_i/\mu s$	$t_o/\mu s$	I/A	U/V	$V_f/$(cm/s)	MRR /(mm^3/min)	REWR/%	SR/μm
1	500	500	5	125	20	8.5953	240.1905	0.0718
2	500	1000	10	150	17.5	16.0266	125.2723	2.4296
3	500	1500	15	175	15	16.7883	2.7661	4.3420
4	500	2000	20	200	12.5	24.2666	23.5356	3.5790
5	500	2500	25	225	10	21.3912	4.7180	5.1562
6	200	500	10	175	12.5	17.5610	22.9198	3.1898
7	200	1000	15	200	10	25.2544	57.6168	3.4744
8	200	1500	20	225	20	27.1082	46.9560	5.3726
9	200	2000	25	125	17.5	12.2999	48.9700	3.0354
10	200	2500	5	150	15	3.7738	421.1362	0.1084
11	100	500	15	225	17.5	33.5894	32.6287	4.7700
12	100	1000	20	125	15	18.2755	41.0075	2.3318
13	100	1500	25	150	12.5	20.2446	5.7735	3.8016
14	100	2000	5	175	10	4.2902	275.0305	0.5402
15	100	2500	10	200	20	7.7399	246.2807	0.8928
16	50	500	20	150	10	26.3527	10.1973	4.1330
17	50	1000	25	175	20	30.3648	43.9864	3.7188
18	50	1500	5	200	17.5	6.8653	207.2457	1.3534
19	50	2000	10	225	15	11.3273	37.1499	2.2564
20	50	2500	15	125	12.5	6.1918	332.7138	0.4580
21	40	500	25	200	15	31.9430	7.4307	4.8824
22	40	1000	5	225	12.5	8.3551	104.7168	3.0976
23	40	1500	10	125	10	6.3368	85.2809	0.1886
24	40	2000	15	150	20	10.1911	90.4258	2.4366
25	40	2500	20	175	17.5	9.2501	189.6518	1.1624

6.4.1 Analysis of MRR

The purpose of the ANOVA is to investigate which machining parameters significantly affect the performance characteristics. In this investigation, the ANOVA and F-test are applied to analyze the experimental data. The contribution of the machining parameters is defined as significant if the calculated F_A values exceed $F_{0.01, n_1, n_2}$.

During the initial ANOVA and F-test associated with the MRR obtained from Table 6.2, MS_{yf} (2.2522) is the minimal and it is less than MS_{Error} (24.7905), so the MS_{yf} is combined into the Error, the final ANOVA and F-test for the MRR is depicted in Table 6.3. As shown in Table 6.3, the peak current (I) and the pulse interval (t_o) are the significant parameters affecting MRR. More electrical energy is conducted into the machining zone within a single pulse upon increasing the peak current. In addition, the electrical discharge frequency increases with a decrease in pulse interval, and more materials could be removed in a unit time. Therefore, the MRR is high with the large peak current and the small pulse interval.

Fig. 6.4 shows the effects of machining parameters on the average MRR. The average MRR is the mean of the MRR for the machining parameter's level. The average MRR_i is the calculated value of y_{Ai}① divided by N②. As shown in Fig. 6.4, the MRR increases with an increase in pulse duration, peak current and peak voltage, respectively. The MRR decreases with the increase of pulse interval. The MRR changes a little with an increase in feed rate of the workpiece. There are many reasons causing these phenomena. A single pulse energy increases with an increase in pulse duration, peak current and peak voltage, respectively. The material removed by a single pulse increases with an increase in a single pulse energy; therefore, the MRR increases. The electrical discharge frequency decreases with an increase in pulse interval, and less material is removed in a unit time; therefore, the MRR decreases. When the feed rate of the workpiece increases, the relative rate between the workpiece and the tool increases, which reduces the energy density of discharge spot on the workpiece, but enhances the mechanical grinding function; therefore, the MRR changes a little. The optimum machining parameters of the compound process for MRR obtained from Table 6.3 and Fig. 6.4 are: 25A peak current, 500μs pulse interval, 225V peak voltage, 500μs pulse duration, 20cm/s feed rate of the workpiece.

---

① y_{Ai} is the sum of ith level of parameter $A(i = 1, 2, 3, 4, 5)$.
② N is the repeating number of each level of parameter A.

Table 6.3 The Final ANOVA and F-test for the MRR

Parameter(A)	Sum of squares (SS_A)	Degree(df_A)	Mean square (MS_A)	F_A	$F_{0.01, n_1, n_2}$
t_i	58.8618	4	14.7155	1.0883	7.01
t_o	619.3954	4	154.8489	11.4522*	7.01
I	969.7491	4	242.4373	17.9299*	7.01
U	308.7952	4	77.1988	5.7094	7.01
Error$_{e+vf}$	108.1707	8	13.5213		
Total	2064.9722	24			

Note: MS_A, the mean square of parameter A; MS_{Error}, the mean square of error. F_A, the F-test value of parameter A. $F_{0.01, n_1, n_2}$, the critical value, and it is quoted from the "Tables for Statisticians".

* Significant parameter

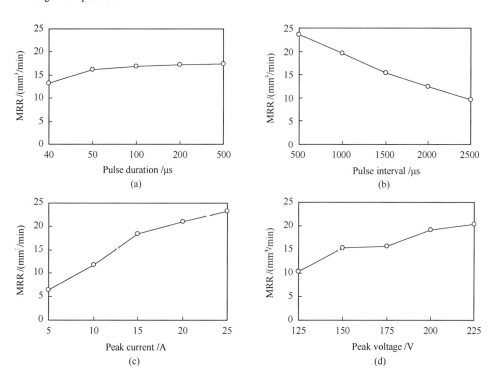

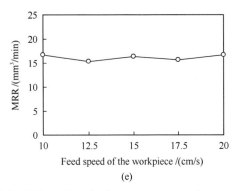

(e)

Fig. 6.4 Effects of machining parameters on the average MRR

In this chapter, the full model for the MRR, REWR and SR are as follows:

$$y = a_0 + \sum_{i=1}^{5} a_i x_i + \sum_{i=1}^{5}\sum_{j=1}^{i} b_{ij} x_i x_j + \sum_{i=1}^{5}\sum_{j=1}^{i}\sum_{k=1}^{j} c_{ijk} x_i x_j x_k \quad (6.2)$$

Where, y= MRR, REWR or SR, $x_1=t_i$, $x_2=t_o$, $x_3=I$, $x_4=U$ and $x_5=V_f$, a_0, a_i, b_{ij}, c_{ijk} is the regression coefficient. Partial F criterion is used to shrink the model.

The capability of the stepwise model to represent the experimental data is assessed through the analysis of variance. The results of the ANOVA assessment for the stepwise regression model of MRR are shown in Table 6.4. As shown in Table 6.4, F is far greater than $F_{0.001,5,29}$, so the relation equation between the MRR and machining parameters can be obtained with stepwise regression analysis method. The equation is as follows:

$$\text{MRR}=10.09664+0.00006909 t_i I^2 - 0.0000000692831 t_o U^2 - 0.000000432393 t_i t_o V_f$$
$$-0.00002844 t_o I V_f + 0.00044353 I U V_f \quad (6.3)$$

Table 6.4 Results of the ANOVA assessment for the stepwise regression model of MRR

Source	DF	Sum of squares	Mean square	F value	$F_{0.001,5,29}$
Model	5	31984	6396.8029	435.22	5.59
Error	29	426.2351	14.6978		
Corrected Total	34	32410.2351			

Fig. 6.5 shows the estimated response surface of the MRR versus peak current and pulse interval, for pulse duration of 500μs, peak voltage of 225V, and feed rate of the workpiece of 20cm/s. It can be seen from this figure that the MRR tends to increase with an increase in peak current for any value of the pulse interval, and increase with the

decrease in pulse interval for any value of the peak current. Furthermore, the maximal MRR can be obtained with the maximal peak current and the minimal pulse interval.

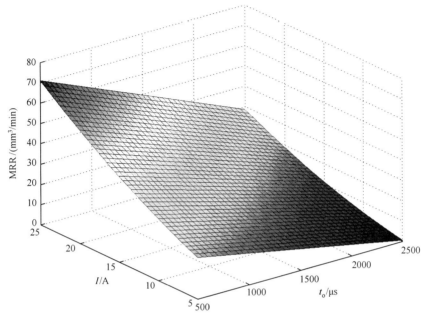

Fig. 6.5 Estimated response surface of the MRR versus I and t_o

6.4.2 Analysis of REWR

During the initial ANOVA and F-test associated with REWR obtained from Table 6.2, MS_{vf} (1767.7134) is the minimal and it is less than MS_{Error} (5602.1018), so the MS_{vf} is combined into the Error, the final ANOVA and F-test for REWR are depicted in Table 6.5. As shown in Table 6.5, the peak current and the pulse interval are the significant parameters affecting REWR.

Table 6.5 Final ANOVA and F-test for REWR

Parameter	Sum of squares	Degree	Mean square	F_A	$F_{0.01, n_1, n_2}$
t_i	7967.8470	4	1991.9617	0.5406	7.01
t_o	109763.9342	4	27440.9836	7.4469*	7.01
I	147848.6515	4	36962.1629	10.0307*	7.01
U	30916.0023	4	7729.0006	2.0975	7.01
$Error_{e+vf}$	29479.2608	8	3684.9076		
Total	325975.6958	24			

* Significant parameter

Fig 6.6 shows the effects of machining parameters on the average REWR. The average REWR is the mean of the REWR for the machining parameter's level. The average $REWR_i$ is the calculated value of y_{Ai} divided by N. As shown in Fig. 6.6, the REWR increases with an increase in pulse interval and feed speed of the workpiece, respectively, and it decreases with an increase in pulse duration, peak current, and peak voltage, respectively. The phenomena can be explained as follows: As shown in Fig. 6.4, the MRR increases with an increase in pulse duration, peak current, and peak voltage, respectively, the thickness of the modified surface layer formed by electrical discharge milling increases with an increase in pulse duration, peak current, and peak voltage, respectively, the grinding force of the tool decreases, and the tool wear is low; therefore, the REWR decreases. The thickness of the modified surface layer formed by electrical discharge milling decreases with the increase of pulse interval and feed rate of the workpiece, the grinding force of the tool increases, the tool wear is high; therefore, the REWR increases. The optimum machining parameters of the compound process for the REWR obtained from Table 6.5 and Fig. 6.6 are: 25A peak current, 500μs pulse interval, 225V peak voltage, 500μs pulse duration, 10cm/s feed rate of the workpiece.

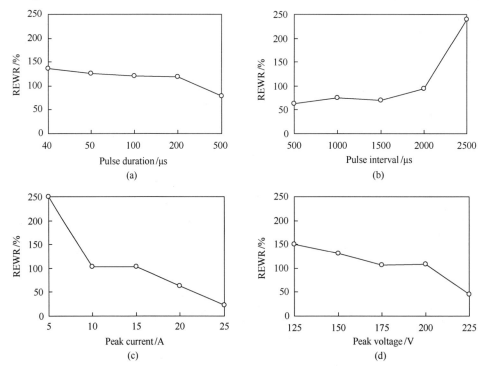

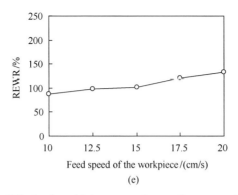

Fig. 6.6   Effects of machining parameters on the average REWR

The results of the ANOVA assessment for the stepwise regression model of the REWR are shown in Table 6.6. As shown in Table 6.6, F is greater than $F_{0.001,9,16}$, so the relation equation between the REWR and machining parameters can be obtained by stepwise regression analysis method. The equation is as follows:

$$\begin{aligned}\text{REWR} =\ & 63.7381 + 0.001177 t_\text{o} U - 0.076357 IU - 0.349 IV_\text{f} + 5.18773 \times 10^{-8} t_\text{o}^3 \\ & - 1.42817 \times 10^{-6} It_\text{o}^2 - 7.79401 \times 10^{-7} t_\text{o}^2 U + 0.002504 I^2 U - 0.000219 t_\text{o} V_\text{f}^2 \\ & + 0.028164 V_\text{f}^3 \end{aligned}$$

(6.4)

Table 6.6   Results of the ANOVA assessment for the stepwise regression model of the REWR

Source	DF	Sum of squares	Mean square	F value	$F_{0.001,9,16}$
Model	9	317499	35278	29.35	5.98
Error	16	19233	1202		
Corrected Total	25	336732			

Fig. 6.7 shows the estimated response surface of the REWR versus peak current and pulse interval, for pulse duration of 500μs, peak voltage of 225V, and feed rate of the workpiece of 10cm/s. The REWR decreases with the increase in the peak current for any value of the pulse interval, and increases with the increase in pulse interval for any value of the peak current. Furthermore, the minimal REWR can be obtained with the maximal peak current and the minimal pulse interval.

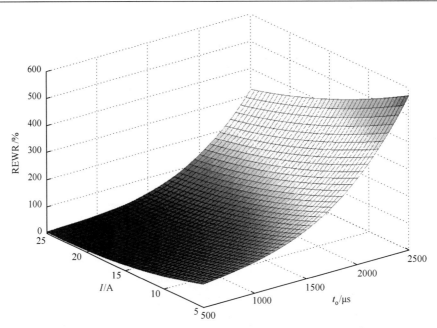

Fig. 6.7 Estimated response surface of the REWR versus I and t_o

6.4.3 Analysis of the SR

During the initial ANOVA and F-test associated with the SR obtained from Table 6.2, MS_{vf} (0.1021) is the minimal and it is less than MS_{Error} (0.6815), so the MS_{vf} is combined into the Error and the final ANOVA and F-test for SR is depicted in Table 6.7. The peak current (I) and the peak voltage (U) are the significant parameters affecting the SR. When the peak current and peak voltage are set at a high level, a huge discharge energy would be delivered into the machining zone within a single pulse, so the machined surface presented a larger and deeper crater size. Therefore, the SR is high with the large peak current and the high peak voltage.

Table 6.7 Final ANOVA and F-test for SR

Parameter	Sum of squares	Degree	Mean square	F_A	$F_{0.01, n_1, n_2}$
t_i	2.7790	4	0.6947	1.7731	7.01
t_o	10.5579	4	2.6395	6.7364	7.01
I	30.7267	4	7.6817	19.6049*	7.01
U	21.4292	4	5.3573	13.6727*	7.01
Error$_{e+vf}$	3.1346	8	0.3918		
Total	68.6274	24			

* Significant parameter

Fig 6.8 shows the effects of machining parameters on average SR. The average SR means the mean of the SR for the machining parameter's level. The average SR_i is the calculated value of y_{Ai} divided by N. The SR increases with the increase in pulse duration, peak current, and peak voltage, respectively, and it decreases with the increase in pulse interval and feed rate of the workpiece, respectively. There are many reasons causing these phenomena. When the pulse duration, peak current and peak voltage increase, the discharge crater generated by a single pulse on the machined surface is deeper and larger because more discharge energy is delivered into the machining zone; therefore, the SR increases. With a longer pulse interval, there is more time to grind the modified surface layer on the workpiece and clear the disintegrated particles from the gap between the tool electrode and the workpiece; therefore, the SR decreases with an increase in pulse interval. When the feed rate of the workpiece increases, the time of repeat electrical discharge at a certain place decreases, and the mechanical grinding function enhances; therefore, the SR decreases. The optimum machining parameters of the compound process for the SR obtained from Table 6.7 and Fig. 6.8 are: 5A peak current, 125V peak voltage, 2500μs pulse interval, 40μs pulse duration, 20cm/s feed rate of the workpiece.

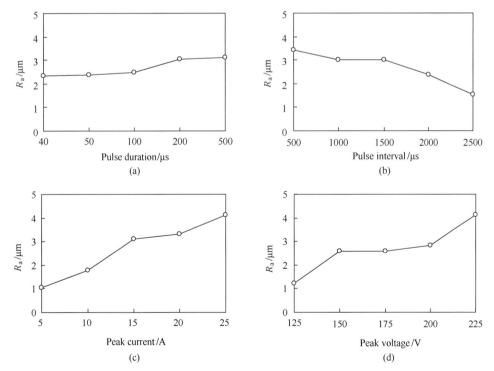

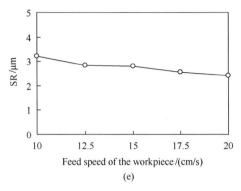

Fig. 6.8 Effects of machining parameters on average the SR

The results of the ANOVA assessment for the stepwise regression model of the SR are shown in Table 6.8. As shown in Table 8, F is greater than $F_{0.001,6,27}$, so the relation equation between the SR and machining parameters can be obtained by stepwise regression analysis method. The equation is as follow:

$$SR = 0.61892 + 0.00080269 t_i V_f + 0.00152 IU - 1.7508 \times 10^{-10} t_o^3 - 0.0000243 6 I^2 U \\ - 0.00004447 t_i V_f^2 + 0.00000137 t_o V_f^2$$

(6.5)

Table 6.8 Results of the ANOVA assessment for the stepwise regression model of the SR

Source	DF	Sum of squares	Mean square	F value	$F_{0.001,6,27}$
Model	6	105.8495	17.6416	44.31	5.31
Error	27	10.7499	0.3981		
Corrected Total	33	116.5994			

Fig. 6.9 shows the estimated response surface of the SR versus peak current and peak voltage, for pulse duration of 40μs, pulse interval of 2500μs, and feed rate of the workpiece of 20cm/s. The SR increases with an increase in peak current for any value of the peak voltage, and increases with an increase in peak voltage for any value of the peak current. Furthermore, the minimal SR can be obtained with the minimal peak current and the minimal peak voltage.

6.4.4 Experiment validation

The optimal combination level of machining parameters has been determined in the previous analysis. The final step is to predict and verify the improvement of the observed values using the optimal combination level of the machining parameters. In this chapter, after determining the optimum conditions and predicting the response

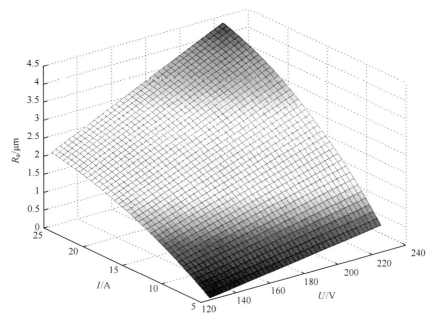

Fig. 6.9 Estimated response surface of the SR versus I and U

under these conditions, a new experiment is designed and conducted with the optimum levels of the machining parameters. Table 6.9 shows the results of the experiment validation for the compound process. The experimental values agree well with predictions, and they are improved significantly in comparison with the experimental values from Table 6.2, which confirms the validity of the used Taguchi method for enhancing the machining performance and optimizing the machining parameters. The photograph of SiC ceramic workpiece of finish machining with the compound process of ED milling and mechanical grinding is shown in Fig. 6.10.

Fig. 6.10 Photograph of finish machining surface of SiC ceramic produced by the compound process

Table 6.9 Results of the confirmation experiment for combined process

	Optimal machining parameters					Predicted	Experimental	Relative error/%
	$t_i/\mu s$	$t_o/\mu s$	I/A	U/V	$V_f/(cm/s)$			
MRR/(mm³/min)	500	500	25	225	20	70.5587	72.5602	2.7584
REWR/%	500	500	25	225	10	2.4487	2.3796	2.9038
SR/μm	40	2500	5	125	20	0.0578	0.0540	7.0370

6.5 Surface integrity of SiC ceramic machined by the compound process

6.5.1 Surface topography of the machined workpiece

The SEM micrographs of the machined surface obtained at three machining conditions with different tool polarities are shown in Fig. 6.11. The surface roughness increases from finish machining to rough machining. This is because a single pulse energy increases from finish machining to rough machining, the crater size increases with an increase in a single pulse energy; therefore, the SR is high. As shown in Fig. 6.11(a)~ Fig. 6.11(d), the surfaces machined at rough machining mode and semi-finish machining mode are characterized by an uneven fusing structure, globules of debris, shallow craters and micropores. The formation of the craters on these surfaces is due to sparks that form at the conductive phase generating melting or possible evaporation. Some of the molten material cools rapidly under the effects of the dielectric fluid, which forms globules on the surface. The formation of the micropores is due to entrapped gases that escape from the solidified material. During rough machining and semi-finish machining, the material is mainly removed by the electrical discharge milling. As shown in Fig. 6.11(e) and Fig. 6.11(f), the surface machined at finish machining mode is smooth, and covered by less little craters and pockmarks. Furthermore, it can be found that there are evident grinding traces on the workpiece surface, which means that the material is mainly removed by mechanical grinding. It can be seen from Fig. 6.11 that the surface appears to be less rough in case of positive tool polarity, when compared to negative tool polarity under the same conditions.

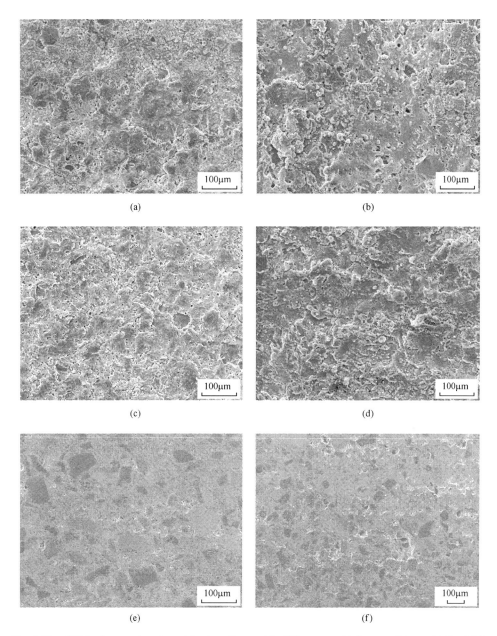

Fig. 6.11 SEM micrographs of machined surfaces with different machining conditions. (a) Tool (+), rough machining, (b) Tool (−), rough machining, (c) Tool (+), semi-finish machining, (d) Tool (−), semi-finish machining, (e) Tool (+), finish machining, (f) Tool (−), finish machining

Higher magnification SEM views of microstructure created by rough machining mode and semi-finish machining mode are shown in Fig. 6.12. There are some micro-cracks on the surface machined at rough machining mode and semi-finish machining mode. The number and length of the micro-cracks on the surface machined by rough machining are higher than that by semi-finish machining. The rapid heating and cooling effect in electric discharge milling induces a high-temperature gradient within the heat affected area and generates a significant stress within the machined surface. When the degree of induced stress exceeds the maximum tensile strength of the SiC ceramic, cracking occurs on the machined surface.

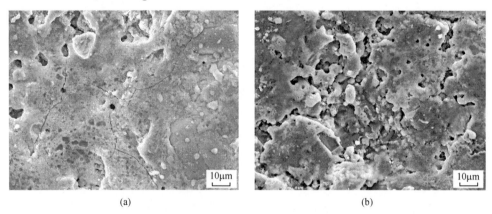

Fig. 6.12　SEM micrographs showing micro-cracks on the machined surfaces with different machining conditions. (a) Tool (+), rough machining, (b) Tool (+), semi-finish machining

6.5.2　Compositions of the machined workpiece

During electric discharge machining, the material can be transferred between the electrodes in solid, molten or gaseous state simultaneously [22,23]. Fig. 6.13~ Fig. 6.15 show the EDS spectrum analysis of the unprocessed surface, and the machined surface with positive tool polarity, and negative tool polarity, respectively. The elements of the specimen are indicated by the peaks corresponding to their energy levels. It can be seen from Fig. 6.13~ Fig. 6.15 that Fe is detected on the machined surface with positive tool polarity, not on the machined surface with negative tool polarity and the unprocessed surface. This implies that some of Fe from the tool electrode diffuses into the specimen surface during the electrical discharge milling with positive tool polarity, whereas the tool electrode material can't transfer to the workpiece surface during the electrical discharge milling with the negative tool polarity.

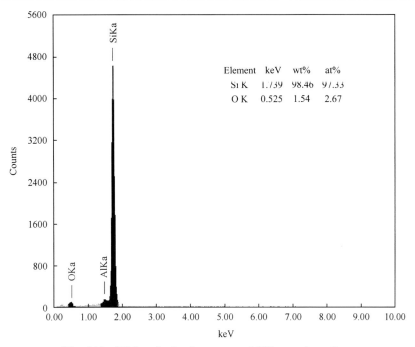

Fig. 6.13 EDS analysis of unprocessed SiC ceramic surface

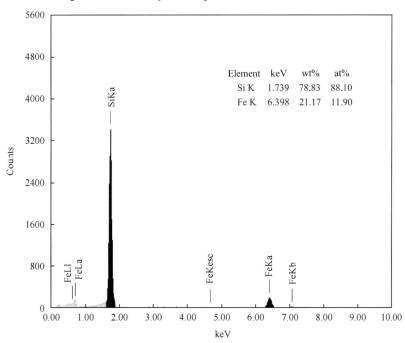

Fig. 6.14 EDS analysis of SiC ceramic surface created at rough machining mode with positive tool polarity

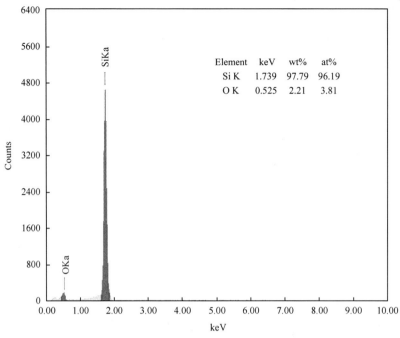

Fig. 6.15 EDS analysis of SiC ceramic surface created at rough machining mode with negative tool polarity

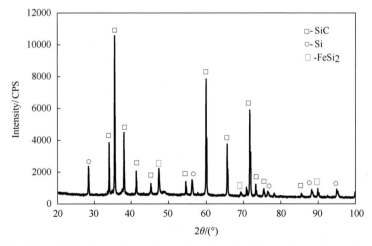

Fig. 6.16 X-ray diffraction patterns of rough machined surface with positive tool polarity

In order to determine the phases of iron or iron compounds, X-ray diffraction pattern of the machined surface at the rough machining mode with positive tool polarity is shown in Fig. 6.16. For comparison, the patterns of the workpiece surface in rough machining with negative tool polarity and unprocessed SiC ceramic are also shown in Fig. 6.17 and Fig. 6.18, respectively. The peaks of FeSi$_2$ indicate that a modified layer forms on the machined surface, and a combination reaction takes place during the compound process when the rough machining mode is used with positive tool polarity, which can be described as follows:

$$Fe + 2Si \longrightarrow FeSi_2$$

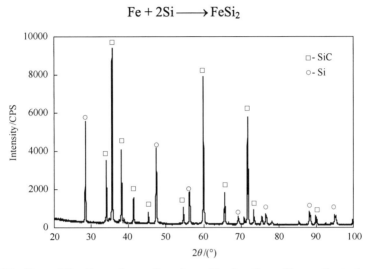

Fig. 6.17 X-ray diffraction patterns of rough machined surface with negative tool polarity

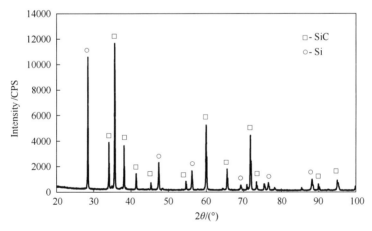

Fig. 6.18 X-ray diffraction patterns of unprocessed machined surface

In order to determine the thickness of the modified layer, the elemental concentration of Fe measured from a line scan of EDS on a cross-section of the rough machined surface with positive tool polarity is shown in Fig. 6.19. The figure shows that the thickness of the modified layer is about 4μm. It can also be seen from Fig. 6.17

and Fig. 6.18 that there is little difference in material composition between the unprocessed surface and the machined surface with negative tool polarity, which means that no evident reaction happens during machining with negative tool polarity. Because the content of oxygen is very few, there is no diffraction peak of oxygen compounds in Fig. 6.17 and Fig. 6.18.

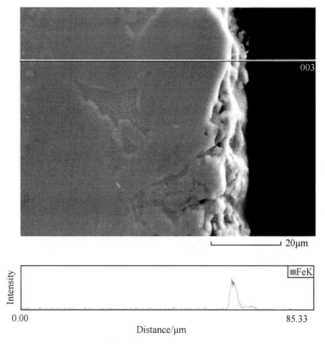

Fig. 6.19　Fe distribution on cross-section obtained by EDS

6.6　Conclusions

The compound process is able to effectively machine a large surface area on SiC ceramics with a good surface quality. Using a water-based emulsion as the machining fluid, harmful gas is not generated during machining, and it shows a good working environmental practice.

　　The peak current and pulse interval are the significant parameters affecting MRR of the compound process. The optimal combination levels of machining parameters with maximized MRR of machining SiC ceramics are: 25A peak current; 500μs pulse interval; 225V peak voltage; 500μs pulse duration; 20cm/s feed rate of the workpiece.

　　The peak current and pulse interval are the significant parameters affecting

REWR of the compound process. The optimal combination levels of machining parameters with minimum the REWR of machining SiC ceramics are: 25A peak current; 500μs pulse interval; 225V peak voltage; 500μs pulse duration; 10cm/s feed rate of the workpiece.

The peak current and peak voltage are the significant parameters affecting the SR of the compound process. The optimal combination levels of machining parameters with minimum the SR of machining SiC ceramics are: 5A peak current; 125V peak voltage; 2500μs pulse interval; 40μs pulse duration; 20cm/s feed rate of the workpiece.

With each optimal combination level of machining parameters, the MRR, REWR and the SR can reach 72.5602mm^3/min, 2.3796% and 0.0540μm, respectively. The experimental values agree well with predictions, and they are improved significantly in comparison with the experimental values obtained at the initial experimental conditions. The experimental results confirm the validity of the used Taguchi method for enhancing the machining performance and optimizing the machining parameters.

The material is mainly removed by the electrical discharge milling when the rough machining mode and the semi-finish machining mode are used, whereas the material is mainly removed by mechanical grinding when finish machining mode is used. There are some micro-cracks on the surface machined at rough machining mode and semi-finish machining mode. The number and length of the micro-cracks on the surface machined by rough machining are higher than that by semi-finish machining.

The material from the tool can transfer to the workpiece and a combination reaction takes place during machining wit positive tool polarity, which forms the thickness of modified layer about 4μm on the machined surface, whereas there is no evident reaction happening during machining with negative tool polarity.

References

[1] Chen M H, Gao L, Zhou J H, et al. Application of reaction sintering to the manufacturing of a spacecraft combustion chamber of SiC ceramics. Journal of Materials Processing Technology, 2002, 129(1-3): 408-411.

[2] Deason D M, Hilmas G, Buchheit A, et al. Silicon carbide ceramics for aerospace applications processing, microstructure, and property assessments. Materials Science and Technology, 2005, 1: 45-53.

[3] Mercurio S, Haber R. Silicon carbide microstructure improvements for armor applications. Ceramic Engineering and Science Proceedings, 2008, 28 (5): 155-159.

[4] Rajgopal S, Zula D, Garverick S, et al. A silicon carbide accelerometer for extreme environment applications. Materials Science Forum, 2007, 600: 859-862.

[5] Zhang Z Y, Yan J W, Kuriyagawa T. Machinability investigation of reaction-bonded silicon carbide by single-point diamond turning. Key Engineering Materials, 2009, 389/390: 151-156.

[6] Puertas I, Luis C J, Villa G. Spacing roughness parameters study on the EDM of silicon carbide. Journal of Materials Processing Technology, 2005, 164/165: 1590-1596.

[7] Luis C J, Puertas I, Villa G. Material removal rate and electrode wear study on the EDM of silicon carbide. Journal of Materials Processing Technology, 2005, 164-165: 889-896.

[8] Yamaguchi S, Noro T, Takahashi H, et al. Electric discharge machining for silicon carbide and related materials. Materials Science Forum, 2009, 600-603: 851-854.

[9] Liu Y H, Ji R J, Li Q Y, et al. An experimental investigation for electric discharge milling of SiC ceramics with high electrical resistivity. Journal of Alloys and Compounds, 2009, 472(1/2): 406-410.

[10] Liu Y H, Ji R J, Li Q Y, et al. Electric discharge milling of silicon carbide ceramic with high electrical resistivity. International Journal of Machine tools and Manufacture, 2008, 48(12/13): 1504-1508.

[11] Ji R J, Liu Y H, Yu L L, et al. Study on high efficient electric discharge milling of silicon carbide ceramic with high resistivity. Chinese Science Bulletin, 2008, 53(20): 3247-3254.

[12] Gopal A V, Rao P V. The optimisation of the grinding of silicon carbide with diamond wheels using genetic algorithms. International Journal of Advanced Manufacturing Technology, 2003, 22(7/8): 475-480.

[13] Zhang Z Y, Yan J W, Kuriyagawa T. Machinability investigation of reaction-bonded silicon carbide by single-point diamond turning. Key Engineering Materials, 2009, 389/390: 151-156.

[14] Gopal A V, Rao P V. A new chip-thickness model for performance assessment of silicon carbide grinding. International Journal of Advanced Manufacturing Technology, 2004, 24(11/12): 816-820.

[15] Churi N J, Pei Z J, Shorter D C, et al. Rotary ultrasonic machining of silicon carbide: Designed experiments. International Journal of Manufacturing Technology and Management, 2007, 12(1/3): 284-298.

[16] Ramulu M. Ultrasonic machining effects on the surface finish and strength of silicon carbide ceramics. International Journal of Manufacturing Technology and Management, 2005, 7(2-4): 107-126.

[17] Sano Y, Watanabe M, Yamamura K, et al. Polishing characteristics of silicon carbide by plasma chemical vaporization machining. Japanese Journal of Applied Physics, Part 1: Regular Papers and Short Notes and Review Papers, 2006, 45(10B): 8277-8280.

[18] Kato T, Sano Y, Hara H, et al. Beveling of silicon carbide wafer by plasma chemical vaporization machining. Materials Science Forum, 2009, 600-603: 843-846.

[19] Samant A N, Daniel C, Chand R H, et al. Computational approach to photonic drilling of silicon carbide. International Journal of Advanced Manufacturing Technology, 2009, 45(7/8): 704-713.

[20] Samant A N, Dahotre N B. Differences in physical phenomena governing laser machining of structural ceramics. Ceramics International, 2009, 35(5): 2093-2097.

[21] Petrofes N F, Gadalla A M. Electrical discharge machining of advanced ceramics. American

Ceramic Society Bulletin, 1988, 67(6): 1048-1052.

[22] Patel K M, Pandey P M, Venkateswara R P. Surface integrity and material removal mechanisms associated with the EDM of Al_2O_3 ceramic composite. International Journal of Refractory Metals and Hard Materials, 2009, 27: 892-899.

[23] Guu Y H, Hou M T K. Effect of machining parameters on surface textures in EDM of Fe-Mn-Al alloy. Materials Science and Engineering A, 2007, 466: 61-67.

Chapter 7　High Speed End Electrical Discharge Milling and Mechanical Grinding of Weakly Conductive Engineering Ceramics

A compound process that integrates end ED milling and mechanical grinding to machine weakly conductive SiC ceramics is developed in this chapter. The process employs a turntable with several uniformly-distributed cylindrical copper electrodes and abrasive sticks as the tool, and uses a water-based emulsion as the machining fluid. End ED milling and mechanical grinding happen alternately and are mutually beneficial, so the process is able to effectively machine a large surface area on SiC ceramic with a good surface quality. The machining principle and characteristics of the technique are introduced. The effects of polarity, pulse duration, pulse interval, open-circuit voltage, discharge current, diamond grit size, emulsion concentration, emulsion flux, milling depth and tool stick number on performance parameters such as the MRR, TWR, and SR have been investigated. Moreover, the effects of pulse duration, pulse interval, discharge current and open voltage on the MRR, the TWR, and the SR have been investigated with Taguchi experimental design. The experimental data are statistically evaluated by analysis of variance and stepwise regression. The significant machining parameters, the optimal combination levels of machining parameters, and the mathematical models associated with the machining characteristics are obtained. The confirmation experiment results confirm the validity of the used Taguchi method for enhancing the machining performance and optimizing the machining parameters. Furthermore, the polarity, pulse duration, and peak current are varied to explore their effects on the surface integrity, such as surface morphology, surface roughness, micro-cracks, and composition on the machined surface.

7.1　Introduction

Over the last few years, there has been a great upsurge of interest in advanced ceramic materials. As a result of this interest, significant advances in the development and use of ceramic materials have been made, and the world markets for advanced ceramics are growing. Of the various ceramic materials, SiC ceramics are the most interested engineering ceramics because of their combination of outstanding mechanical, physical and chemical properties such as low density, high strength, high thermal conductivity, low thermal expansion coefficient and wear and high corrosion resistance even at elevated temperature [1-3]. The broad range of technological applications presently served by SiC ceramics includes cutting tools, automotive engine parts, heat exchangers, high temperature bearings, heavy-duty electric contacts, fixtures, nozzles, turbine blades and many more applications [4-6]. However, the properties that make these materials appealing to use also create a major challenge in traditional diamond grinding or diamond turning, because of their great hardness and brittleness [7,8].

The EDM is a thermoelectric process, whereby material is removed by a succession of electrical discharges occurring between an electrode and a workpiece which is immersed in a dielectric liquid medium. Unlike traditional cutting and grinding processes, which rely on the force generated by a harder tool or abrasive material to remove the softer work-material, the EDM process utilizes electrical sparks or thermal energy to erode the unwanted material and generate the desired shape. Since no mechanical contact between the electrode and the workpiece is involved during the EDM, the process is able to machine any conductive component into accurate and complex shapes irrespective of its high hardness and strength [9-11].

Since the EDM has been shown to be an effective and economical technique for machining difficult-to-machine materials, it is believed that the EDM process will open up an opportunity for the machining of SiC ceramics. However, conventional EDM techniques such as die-sinking electrical discharge machining and WEDM show low efficiency when machining a large surface area on a SiC ceramic [12,13]. EDG integrates electrical discharge machining and mechanical grinding. During this process, the electrical discharge and grinding processes occur simultaneously, so electrically conductive hard materials can be machined with good surface quality, but the material removal rate is low, and the cost is high [14, 15]. Recently, some progress has been made to improved efficiency by using EDM milling. The MRR can be improved significantly, but it still cannot meet the demand of modern industrial applications, and the machined surface is poor [16-19].

A novel high speed compound machining process that integrates end ED milling and mechanical grinding to machine SiC ceramic is proposed in this chapter. A L_{25} orthogonal array based on Taguchi method is used to investigate the effects of pulse duration, pulse interval, discharge current and open voltage on the machining characteristics. The experimental data are statistically analyzed by the ANOVA and F-test, and a series of mathematical models associated with the MRR, the TWR and the SR are established with the stepwise regression method.

7.2 Principle and characteristics for end ED milling and mechanical grinding of SiC ceramic

7.2.1 Principles

The principle of end ED milling and mechanical grinding of SiC ceramic is shown in Fig. 7.1. The tool and the workpiece are connected to the negative and positive poles of the pulse generator, respectively. The tool is a turntable with several uniformly-distributed small cylindrical electrodes and abrasive sticks rotating rapidly around its axis. The tool is mounted onto a rotary spindle, driven by an A.C. motor. The workpiece is SiC ceramic blank and is mounted onto a NC table. The machining fluid is a water-based emulsion.

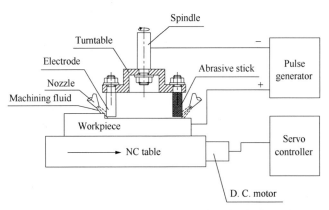

Fig. 7.1 Schematic illustration of end ED milling and mechanical grinding of SiC ceramic

During machining, the tool rotates at a high speed, the machining fluid is flushed into the gap between the tool and the workpiece with several nozzles, and the SiC ceramic workpiece is fed towards the tool by the NC table. As the workpiece approaches the tool and the distance between the workpiece and the electrode reaches the discharge gap, electrical discharges are produced. A plasma channel with high temperature and high pressure grows during the pulse duration [20]. The instantaneous high temperature and pressure plasma causes SiC ceramic to be removed by electrical discharge milling, and a modified surface layer is formed on the workpiece surface. The following abrasive stick grinds the modified surface layer. The discharge energy thermally softens the modified surface layer in the grinding zone, and consequently decreases the normal force and the grinding power, allowing the modified surface layer to be removed easily by the abrasive stick, which can also enhance the effectiveness of the subsequent discharge. The combination of these alternating, mutually beneficial processes means that a high material removal rate and good surface quality can be achieved when machining SiC ceramic.

During this compound machining, the discharge gap will match the machining condition, and the abrasive sticks and the electrodes are worn at the same rate. If the wear of the abrasive stick was greater than that of the electrode, the distance between the abrasive stick and the workpiece surface would increase, weakening the mechanical grinding function so that the electrical discharge function becomes stronger; the material on the workpiece surface is then mostly removed by electrical discharges, which will in turn accelerate the electrode wear, to eventually achieve the same wear of the electrodes and the abrasive sticks. If the wear of the electrode was greater than that of the abrasive stick, the distance between the electrode and the workpiece surface would increase, the electrical discharge function become weaker, the mechanical grinding function become stronger, until the material on the workpiece surface is mostly removed by the mechanical grinding function, which would accelerate the abrasive stick wear, until eventually the same wear of the abrasive sticks and the electrodes was obtained.

Comparing with ED milling and conventional mechanical grinding, the advantages of the compound machining process can be explained as follows: During ED milling of SiC ceramic, a modified surface layer will form on the workpiece surface, which makes the subsequent discharge difficult and degrades the workpiece surface quality. During conventional mechanical grinding of SiC ceramic, the grinding force is high because of the high hardness of the workpiece, so the material removal rate is low. During this compound process, mechanical grinding removes the softened material created by the end electrical discharge milling, to produce higher overall machining performance. The comparison experiments have been carried out for positive tool polarity, pulse duration of 50μs, pulse interval of 2500μs, discharge current of 15A, open-circuit voltage of 90V, diamond grit size of #120 in the abrasive sticks, and diamond concentration of 100% in the abrasive sticks. The experimental results are shown in Fig. 7.2. A higher MRR, lower TWR and lower SR can be obtained with the compound process in comparison with ED milling alone. Furthermore, although the SR with the compound process is a little higher than that with conventional mechanical grinding, the MRR is far higher and the TWR is far lower with the compound process in comparison to conventional mechanical grinding.

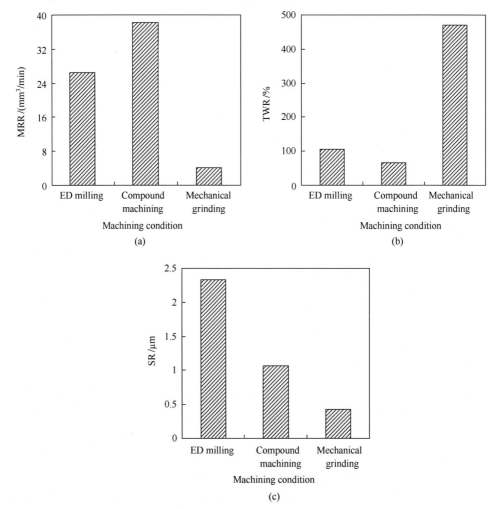

Fig. 7.2 Comparison of experimental results for machining of SiC ceramic with different processes. (a) Effect of machining condition on the MRR, (b) Effect of machining condition on the TWR, (c) Effect of machining condition on the SR

7.2.2 Characteristics of end ED milling and mechanical grinding of SiC ceramic

The advantages of combined end ED milling and mechanical grinding for SiC ceramic are listed below.

(1) A large diameter turntable with several cylindrical copper electrodes and abrasive sticks is used as the tool (Fig. 7.3), so a large surface area can be easily machined by the compound process. The process shows high material removal rate and good surface quality. Fig. 7.4 is a photograph of a SiC ceramic workpiece machined by the compound process.

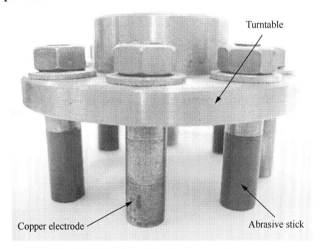

Fig. 7.3 Tool used in the compound process

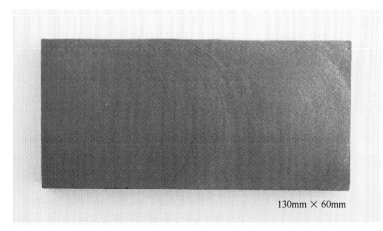

Fig. 7.4 Photograph of SiC ceramic workpiece machined by the compound process

(2) The cylindrical copper electrodes and abrasive sticks used (Fig. 7.5), can be

easily and economically manufactured. Furthermore, the cylindrical copper electrodes and abrasive sticks are easily fixed and replaced, eliminating the need for a high-cost monolithic tool.

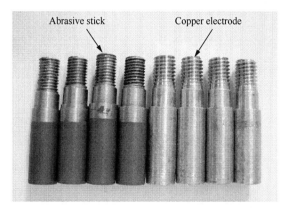

Fig. 7.5　Cylindrical copper electrodes and abrasive sticks used in the compound process

(3) The cylindrical electrodes and abrasive sticks are fixed on the turntable alternately and separately; the machining fluid is flushed into the gap, and the chips are flushed away easily. Therefore, the compound process improves the discharge stability and machining efficiency.

(4) A water-based emulsion is used as the machining fluid, so harmful gas is not generated during the compound process, and the equipment is not corroded.

(5) Rough machining, semi-finish machining and finish machining can be obtained on the same machine by adjusting the discharge parameters when a large surface on a SiC ceramic workpiece is machined.

7.3　Experiments

7.3.1　Experimental procedures

The experiments were carried out in an electrical discharge milling machine. In the following experiments, the workpiece material is SiC ceramic, the tool is a turntable with several uniformly-distributed cylindrical copper electrodes and cast iron bonded diamond abrasive sticks around the circumference. The copper electrodes and abrasive sticks are fixed on the turntable alternately (Fig. 7.3). The diameter of the cylindrical copper electrode is 10mm, the diameter of the abrasive stick is 10mm, the diameter of the turntable is 90mm, the rotational speed of the spindle is 3000r/min. The machining fluid is a water-based emulsion composed of emulsified oil and distilled water, which were mixed with a constant speed power-driver mixer. The MRR and TWR are as defined in Equation (4.1) and Equation (4.2), respectively, and are obtained through measuring the dimensions of the workpiece and the tool stick before and after machining with a dial indicator and a vernier caliper. The SR is measured with a surface roughness tester (TR220, Beijing Time High Technology Ltd., China). The microstructure of the workpiece surface is examined with SEM (JEOL JSM-6380, Japan), equipped with EDS (JEOL JED-2300, Japan). Unless otherwise specified, the following experimental parameters are used: the tool polarity was negative, the pulse duration is 400μs, the pulse interval is 300μs, the open-circuit voltage is 150V, the discharge current is 75A, the grit size of diamond in the abrasive sticks is #120 (grain size = 124μm), the concentration of diamond in the abrasive sticks was 100%, the emulsion concentration is 8wt%, the emulsion flux is 200ml/s, the milling depth is 0.1mm, and the number of tool sticks is 8 (four abrasive sticks and four copper electrodes).

7.3.2 Experimental design

Four machining parameters such as pulse duration, pulse interval, discharge current and open voltage, are chosen to explore their effects on the machining characteristics in terms of the MRR, TWR and SR. The input machining parameters and their levels are presented in Table 7.1. Out of the standard orthogonal arrays available in Taguchi design, a L_{25} orthogonal array is selected for this work.

Table 7.1 Machining parameters and their levels

Factors	Levels				
	1	2	3	4	5
Pulse duration/μs	400	120	90	50	40
Pulse interval/μs	300	600	1200	1800	2500
Discharge current/A	15	30	45	60	75
Open voltage/V	90	110	130	150	170

7.3.3 Analysis and discussion of experimental results

Based on the Taguchi method, the ANOVA and F-test are used to investigate the relative effect of different machining parameters on the machining characteristics, and to determine the optimal combination levels of machining parameters. F-value of the machining parameter is compared with the appropriate confidence table. When F value of the machining parameter is bigger than F_{α, n_1, n_2} value of the confidence table, where α is risk, n_1 and n_2 are degrees of freedom associated with numerator and denominator, the contribution of the machining parameter is defined as significant. During this investigation, there are three categories of significant machining parameters: ① More significant machining parameters (α is 0.001); ② Significant machining parameters (α is 0.01); ③ Less significant machining parameters (α is 0.1). F_{α, n_1, n_2} is quoted from the "Tables for Statisticians" [21].

To obtain applicable and practical predictive quantitative relationships, the mathematical models for the MRR, the TWR, and the SR are obtained by using the stepwise regression method with a commercially available mathematical software SAS (version 8.0).

7.4 Results and discussion of the single factor experiment

7.4.1 Effect of tool polarity on the process performance

Tool polarity is a primary factor that affects process performance parameters such as the MRR, the TWR and the SR. The effect of tool polarity on the process performance is shown in Fig. 7.6.

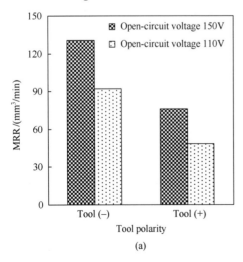
(a)

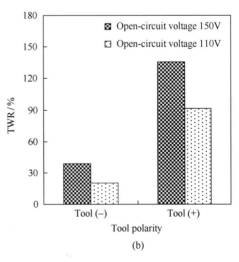
(b)

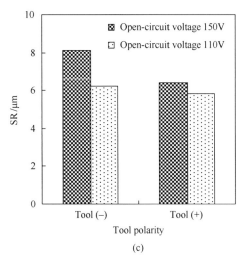

Fig. 7.6 Effect of tool polarity on the process performance. (a) Effect of tool polarity on MRR, (b) Effect of tool polarity on TWR, (c) Effect of tool polarity on SR

The MRR values obtained with different tool polarities are given in Fig. 7.6(a), under the same conditions, the MRR with negative tool polarity is 1.7~1.9 times that with positive tool polarity. The TWR values for different tool polarities are given in Fig. 7.6(b); under the same conditions the TWR for positive tool polarity is 3.5~4.5 times higher than that for negative tool polarity. These phenomena can be explained as follows: The tool rotates at a high speed during machining, the discharge point transfer velocity between the electrode and workpiece is very high, and the lifetime of the discharge between a particular point of the electrode and a particular point of the workpiece is very short. Because the mass of the electrons is much smaller than that of positive ions, and they can be accelerated quickly over a short time, the bombardment effect by electrons is stronger than that by positive ions; the MRR is higher in negative tool polarity, and the TWR is higher in positive tool polarity.

Fig. 7.6(c) shows the influence of tool polarity on the SR; under the same conditions, the SR with negative tool polarity is 1.1~1.3 times higher than that with positive tool polarity. Because the bombardment by positive ions is weaker than that by electrons, the craters on the workpiece surface produced by positive ions are shallow; the SR is lower when the tool polarity is positive.

7.4.2 Effect of pulse duration on the process performance

The effect of pulse duration on the process performance is given in Fig. 7.7. As shown in Fig. 7.7(a), the MRR increases gradually with an increase in pulse duration. This is because the discharge energy delivered into the machining zone within a single pulse increases with an increase in pulse duration. Thermal erosion effects, such as vaporization and melting of the machined surface, are enhanced, and thus the MRR increases.

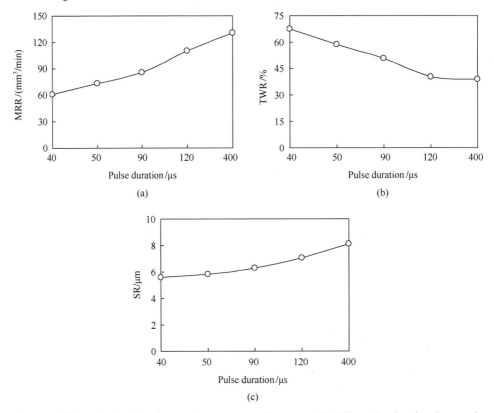

Fig. 7.7 Effect of pulse duration on the process performance. (a) Effect of pulse duration on the MRR, (b) Effect of pulse duration on the TWR, (c) Effect of pulse duration on the SR

Fig. 7.7(b) shows the influence of pulse duration on the TWR; the TWR decreases with increasing pulse duration. This phenomenon can be explained as follows: During EDM, a layer can be deposited on the electrode surface because of the decomposition of the dielectric fluid and workpiece material attached to the tool electrode surface, and the tool wear can be prevented by the protective effects of this deposited layer [22,23]. As the pulse duration increases, the discharge energy delivered to the machining gap increases; the dielectric and workpiece are heated for more time, the carbon released as a result of decomposition of the dielectric is easily attached to the copper electrode surface, resulting in increased deposition on the electrode. Moreover, the thickness of the modified surface layer formed by electrical discharge milling increases with increasing pulse duration; thus the grinding force decreases and the abrasive stick wear is low, therefore, the TWR decreases.

The effect of pulse duration on the SR is shown in Fig. 7.7(c); the SR increases with an increase in pulse duration. This is because the crater size generated by a single pulse becomes larger as the energy of that single pulse increases with the increase in pulse duration.

7.4.3 Effect of pulse interval on the process performance

The effects of pulse interval on the MRR, the TWR and the SR are illustrated in Fig. 7.8(a), Fig. 7.8(b), Fig. 7.8(c), respectively. Fig. 7.8(a) shows that the MRR decreases with an increase in pulse interval. The reason for this is that the electrical discharge frequency and the discharge energy delivered to the machining gap decrease as the pulse interval increases, and decrease the MRR.

As shown in Fig. 7.8(b), the TWR increases with an increase in pulse interval. This phenomenon can be explained as follows: The time for deionization of the dielectric increases with an increase in pulse interval; the discharge energy delivered to the machining gap per unit time decreases and thus the amount carbon released via decomposition of the dielectric decreases, which weaken the deposition effect and increase the grinding force on the abrasive sticks. This results in high abrasive stick wear and the TWR increases.

The effect of pulse interval on the SR is shown in Fig. 7.8(c); the SR initially increases with an increase in pulse interval, and then decreases with further increasing of pulse interval. There are many reasons behind this phenomenon. A longer pulse interval means more time for deionization of the dielectric, which increases the breakdown voltage and the discharge explosion force; the crater size generated by a single pulse becomes larger and deeper, therefore, the SR increases with an increase in pulse interval. However, once the pulse interval is larger than 1200μs, the breakdown voltage and the discharge explosion force do not increase any further, and the crater amount generated by electrical discharge decreases, moreover, there is more time to grind the modified surface layer from the workpiece and clear the disintegrated particles from the gap between the tool and the workpiece; therefore, the SR decreases with further increase of pulse interval.

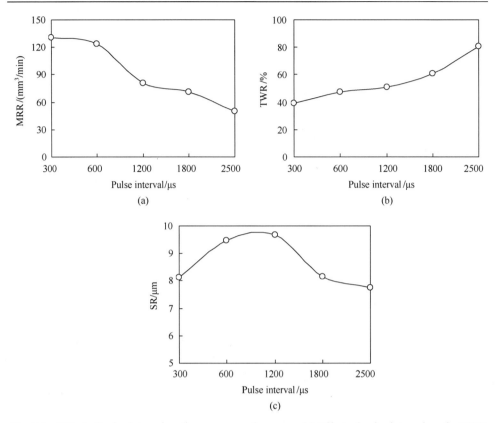

Fig. 7.8 Effect of pulse interval on the process performance. (a) Effect of pulse interval on the MRR, (b) Effect of pulse interval on the TWR, (c) Effect of pulse interval on the SR

7.4.4 Effect of open-circuit voltage on the process performance

The effect of open-circuit voltage on the process performance is shown in Fig. 7.9. Fig. 7.9(a), Fig. 7.9(b) shows that the MRR and the TWR increase with an increase in open-circuit voltage. The phenomenon can be explained as follows: The single pulse energy, thermal energy density and discharge explosive force all increase with an increase in open-circuit voltage, which enhances the removal of material from both the workpiece and the tool, and result in increasing the MRR and the TWR.

Fig. 7.9(c) shows the influence of open-circuit voltage on SR. The SR increases with an increase in open-circuit voltage. The reason for this is that the material removed by a single pulse increases as the open-circuit voltage increases; the discharge crater becomes larger and deeper, and result in increasing the SR.

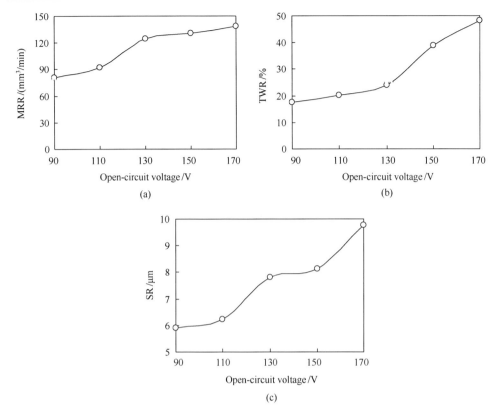

Fig. 7.9 Effect of open-circuit voltage on the process performance. (a) Effect of open-circuit voltage on the MRR, (b) Effect of open-circuit voltage on the TWR, (c) Effect of open-circuit voltage on the SR

7.4.5 Effect of discharge current on the process performance

The effects of discharge current on the MRR, the TWR and the SR are illustrated in Fig. 7.10(a), Fig. 7.10(b) and Fig. 7.10(c), respectively. Fig. 7.10(a) shows that the MRR increases with an increase in discharge current. The phenomenon can be explained by the relationship between the material removed by a single pulse and discharge current [Equation (1.1)]. Under constant pulse duration, the material removed by a single pulse increases with an increase in discharge current; therefore, the MRR increases.

The effect of discharge current on the TWR is shown in Fig. 7.10(b); the TWR increases with an increase in discharge current. This is because the single pulse energy, thermal energy density and discharge explosive force all increase with an increase in discharge current; this enhances the material removal from the tool, so the TWR increases.

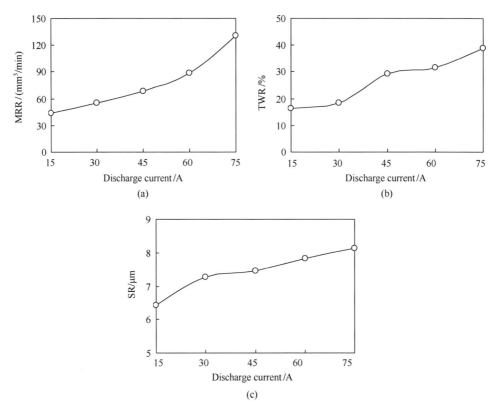

Fig. 7.10 Effect of discharge current on the process performance. (a) Effect of discharge current on the MRR, (b) Effect of discharge current on the TWR, (c) Effect of discharge current on the SR

As shown in Fig. 7.10(c), the SR increases gradually with an increase in discharge current. This is because the crater size generated by a single pulse becomes larger with an increase in single pulse energy. Single pulse energy increases with an increase in discharge current; the SR also increases.

7.4.6 Effect of diamond grit size on the process performance

The effect of diamond grit size on the process performance is illustrated in Fig. 7.11. The MRR with different diamond grit sizes is given in Fig. 7.11(a); under the same conditions, the MRR for #60 diamond grit size is 1.0~1.1 times that with #120 diamond grit size. The TWR with different diamond grit sizes is given in Fig. 7.11(b); under the same conditions the TWR for #60 diamond grit size is 1.2~1.3 times higher than that with #120 diamond grit size. The explanation for this is that as the diamond grit size decreases, the diamond grain size increases, so the grinding force of the abrasive sticks increases, and the removal of the modified surface layer and the removal of the tool sticks are both enhanced. Therefore, the MRR and TWR increase. As shown in Fig. 7.11(c), the SR with #60 diamond grit size is 1.0 to 1.1 times higher than that with #120 diamond grit size. The reason for this is that as the grinding force increases with the decrease in diamond grit size, the grinding trace becomes more evident, and the SR increases.

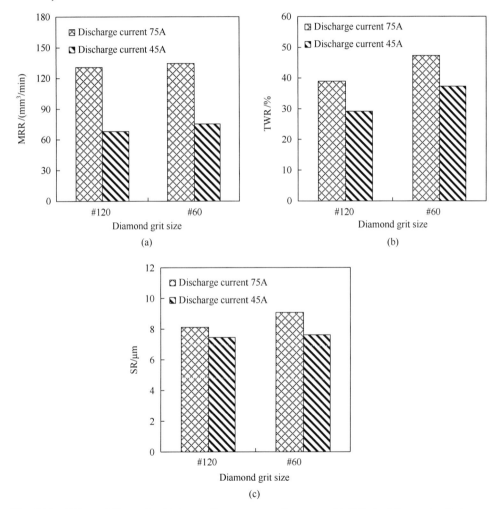

Fig. 7.11 Effect of diamond grit size on the process performance. (a) Effect of diamond grit size on the MRR, (b) Effect of diamond grit size on the TWR, (c) Effect of diamond grit size on the SR

7.4.7 Effect of emulsion concentration on the process performance

The effect of emulsion concentration on the process performance is shown in Fig. 7.12. As shown in Fig. 7.12(a), the MRR initially increases with an increase in emulsion concentration, and then decreases with further increase in emulsion concentration. This is a complex phenomenon. The dielectric strength, washing capability, density and viscosity of the machining fluid increase as the emulsion concentration increases, which enhances the pinch-effect and energy density of the discharge channel, and increase the ejection effect of the eroded material; therefore, the MRR rises. However, once the viscosity of the machining fluid becomes very high, the eroded materials are difficult to flush away; the stability of electrical discharges becomes unsatisfactory, and hence the MRR falls.

Fig. 7.12(b) shows the effect of emulsion concentration on the TWR; the TWR initially decreases with an increase in emulsion concentration, and then increases with further increase in emulsion concentration. The phenomenon can be explained as follows: There are hydrocarbons in the emulsion, and a layer deposits on the electrode surface because of the decomposition of these hydrocarbons during electrical discharges; this layer can prevent the tool wear. As the emulsion concentration increases, the hydrocarbon content in the dielectric increases, the decomposition of the hydrocarbon and the formation of deposits on the electrode surface are enhanced; therefore, the TWR decreases. However, once the emulsion concentration exceeds 12%, the viscosity of the machining fluid increases substantially. The eroded materials become difficult to flush away, and they are gathered in the machining zone. The electrical discharge energy supplied to the machining zone repeatedly strikes the un-expelled eroded materials that are concentrated on the machined surface, which causes unnecessary electrode wear; therefore, the TWR is high.

The effect of emulsion concentration on SR is shown in Fig. 7.12(c); the SR increases with an increase in emulsion concentration. This is because the energy density of the discharge channel increases with an increase in emulsion concentration; the crater size generated by a single pulse becomes larger and deeper, therefore, the SR increases.

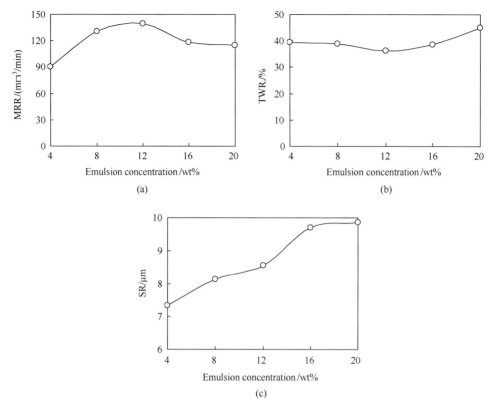

Fig. 7.12 Effect of emulsion concentration on the process performance. (a) Effect of emulsion concentration on the MRR, (b) Effect of emulsion concentration on the TWR, (c) Effect of emulsion concentration on the SR

7.4.8 Effect of emulsion flux on the process performance

The effects of emulsion flux on the MRR, the TWR and the SR are illustrated in Fig. 7.13(a), Fig. 7.13(b) and Fig. 7.13(c), respectively. It can be seen from Fig. 7.13(a) that the MRR increases with an increase in emulsion flux. This phenomenon can be explained as follows: A high emulsion flux results in better cooling of the electrode and better removal of the eroded materials. Under these conditions, the electrical discharges become strong and stable, therefore, the MRR increases.

Fig. 7.13(b) shows the effect of emulsion flux on the TWR; the TWR increases with an increase in emulsion flux. Again, this is because of better cooling of the electrode and better removal of the eroded materials. The deposit formation on the electrode surface decreases, and the mechanical grinding function is enhanced; therefore, the TWR increases.

As shown in Fig. 7.13(c), the SR decreases with an increase in emulsion flux. This phenomenon can be explained as follows: The high emulsion flux enhances electrode cooling and removal of the eroded materials; the craters generated by the electrical discharges are shallow and uniformly distributed on the workpiece surface, and the mechanical grinding function is enhanced, so the SR decreases with an increase in emulsion flux.

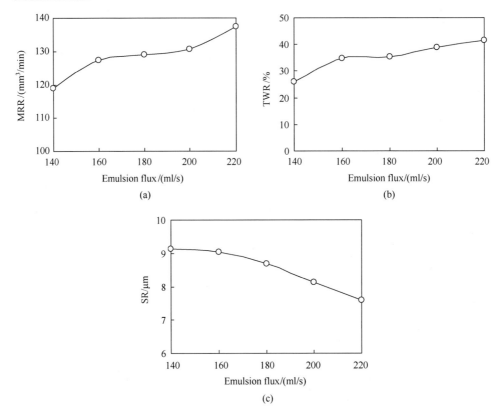

Fig. 7.13 Effect of emulsion flux on the process performance. (a) Effect of emulsion flux on the MRR, (b) Effect of emulsion flux on the TWR, (c) Effect of emulsion flux on the SR

7.4.9 Effect of milling depth on the process performance

The effect of milling depth on the process performance is illustrated in Fig. 7.14. As shown in Fig. 7.14(a), the MRR initially increases with an increase in milling depth and then decreases with further increase of milling depth. This is also a complicated phenomenon. A shallower milling depth leads to higher discharge current density, and thus the thermal energy density of the discharge channel is very high. At this time, the material is removed by vaporization. As the heat consumed by vaporization is high, the MRR is low. However, at a very deep milling depth, the discharge current density is low, which makes it difficult for the SiC ceramic to be removed, and the electrical discharge becomes unstable; therefore, once the milling depth is above 0.1mm, the MRR decreases with further increase in milling depth.

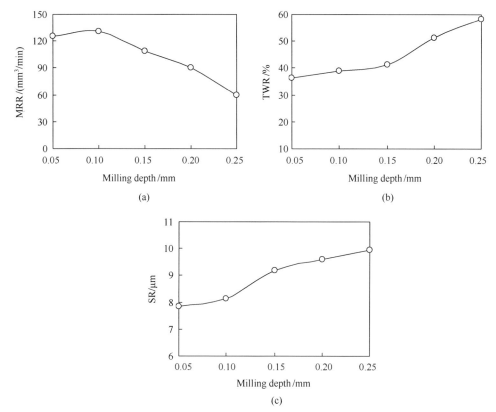

Fig. 7.14　Effect of milling depth on the process performance. (a) Effect of milling depth on the MRR, (b) Effect of milling depth on the TWR, (c) Effect of milling depth on the SR

Fig. 7.14(b), Fig. 7.14(c) show that the TWR and the SR increase with an increase in milling depth. The phenomenon can be explained as follows: The discharge current density decreases with an increase in milling depth. The electrical discharge becomes unstable and arcs are easily generated, which causes unnecessary tool stick wear and deteriorates the machined surface; therefore, the TWR and SR increase.

7.4.10　Effect of tool stick number on the process performance

The effects of tool stick number on the MRR, the TWR and the SR are shown in

Fig. 7.15(a), Fig. 7.15(b), Fig. 7.15(c), respectively. It can be seen from Fig. 7.15(a) and Fig. 7.15(c) that the MRR increases with an increase in tool stick number, while the SR decreases. The phenomena can be explained as follows: Since the tool stick diameter and the turntable diameter are constant, as the tool stick number increases, the total electrical discharge time per unit time increases, and the mechanical grinding function is enhanced; therefore, the MRR increases, and the SR decreases.

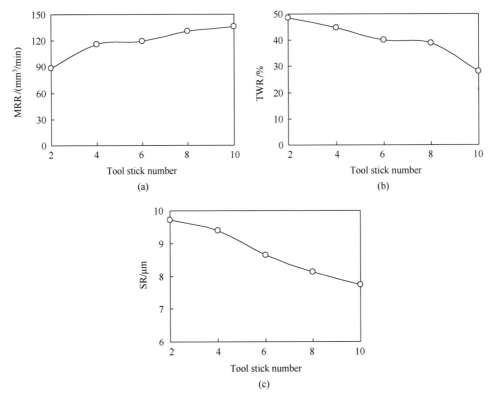

Fig. 7.15 Effect of tool stick number on the process performance. (a) Effect of tool stick number on the MRR, (b) Effect of tool stick number on the TWR, (c) Effect of tool stick number on the SR

The effect of tool stick number on the TWR is shown in Fig. 7.15(b); the TWR decreases with an increase in tool stick number. This phenomenon can be explained as follows: The total electrical discharge time per unit time increases with an increase in tool stick number, and the discharge energy delivered to the machining gap increases. This means that the dielectric and workpiece are heated for more time, and since the carbon released by decomposition from the hydrocarbon in the emulsion is easily attached to the copper electrode surface, the deposition effect is enhanced; moreover, the thickness of the modified surface layer formed by electrical discharge milling increases with an increase in tool stick number, Hence, the grinding force decreases, and the abrasive stick wear is low; the TWR decreases.

7.5 Analysis of Taguchi method

The design matrix from the L_{25} orthogonal array based on the Taguchi method, and the observed values of the MRR, the TWR and the SR are given in Table 7.2.

Table 7.2 Orthogonal array L_{25} and observed values of the MRR, the TWR, and the SR

Number of experiment	$t_i/\mu s$	$t_o/\mu s$	I/A	U/V	MRR/ (mm³/min)	TWR/%	SR/μm
1	400	300	15	90	19.9644	8.1797	5.8002
2	400	600	30	110	53.7118	24.2161	6.9011
3	400	1200	45	130	58.3786	38.8336	7.2325
4	400	1800	60	150	70.1763	59.4613	7.2680
5	400	2500	75	170	63.7853	82.1173	7.8835
6	120	300	30	130	58.5639	34.5533	6.7117
7	120	600	45	150	60.1037	52.1895	6.8064
8	120	1200	60	170	94.3671	61.6708	6.7709
9	120	1800	75	90	36.0003	40.6191	5.3267
10	120	2500	15	110	10.4548	34.5641	4.8149
11	90	300	45	170	101.9575	64.2443	6.7235
12	90	600	60	90	35.9634	46.9493	5.0071
13	90	1200	75	110	36.0096	77.0633	5.3504
14	90	1800	15	130	27.5794	32.8821	3.7405
15	90	2500	30	150	22.8542	74.7907	4.5161
16	50	300	60	110	38.1201	40.7014	5.0899
17	50	600	75	130	74.8590	72.6125	5.1492
18	50	1200	15	150	26.0074	48.9399	3.9891
19	50	1800	30	170	43.5350	81.5669	5.5753
20	50	2500	45	90	8.6003	53.4088	3.8577
21	40	300	75	150	82.3157	77.1034	5.7894
22	40	600	15	170	30.9002	66.6335	4.6521
23	40	1200	30	90	10.9877	52.2245	3.4801
24	40	1800	45	110	22.1535	67.0313	4.0601
25	40	2500	60	130	30.8436	66.8525	4.5809

7.5.1 Analysis of Taguchi method for MRR

The results of the ANOVA and F-test for the MRR are given in Table 7.3, It can be observed from Table 7.3 that open voltage is the significant parameter on the MRR, whereas discharge current and pulse interval are the less significant parameters on the MRR.

Table 7.3 Analysis of variance and F-test for the MRR

Parameter	Sum of squares	Degree	Mean square	F
Pulse duration	1259.109	4	314.7772	1.51
Pulse interval	3033.310	4	758.3275	3.64*
Discharge current	4128.892	4	1032.2230	4.96*
Open voltage	6200.310	4	1550.0780	7.44**
Error	1666.219	8	208.2773	
Total	16287.840	24		

* Less significant parameter
** Significant parameter

The average values of MRR at levels 1, 2, 3, 4, and 5 of the four input machining parameters are calculated to investigate the parameter effects. The effects of machining parameters on average MRR are shown in Fig. 7.16. The MRR increases with an increase in pulse duration, discharge current, and open voltage, respectively, and it decreases with an increase in pulse interval. There are many reasons causing the phenomena. The single pulse energy increases with an increase in pulse duration, discharge current, and open voltage, respectively, and the material removed by a single pulse increases with the single pulse energy increasing; therefore, the MRR increases. In addition, the electrical discharge frequency decreases with an increase in pulse interval, and less material is removed in a unit time, so the MRR decreases with an increase in pulse interval. The optimum machining parameters of the compound process for the MRR obtained from Fig. 7.16 are as follows: 400μs pulse duration; 300μs pulse interval; 75A discharge current; and 170V open voltage.

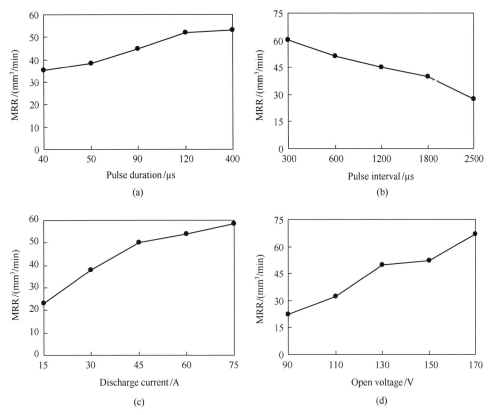

Fig. 7.16 Effects of machining parameters on the average MRR. (a) Effect of pulse duration on the average MRR, (b) Effect of pulse interval on the average MRR, (c) Effect of discharge current on the average MRR, (d) Effect of open voltage on the average MRR

The capability of the stepwise model to represent the experimental data is assessed through the analysis of variance. The results of the analysis of variance for the stepwise model of the MRR are shown in Table 7.4. As in this case F is far greater than $F_{0.001,7,17}$, it is accepted that the mathematical model of MRR by using stepwise regression analysis method is appropriate for a significance level, α, of 0.001. The mathematical relation between the MRR and machining parameters is as follows:

$$\text{MRR} = -0.95554 + 0.00274 t_i U + 1.992879 \times 10^{-7} t_i^2 t_o - 0.00000569 t_i^2 U$$
$$- 4.0673 \times 10^{-8} t_i t_o^2 + 1.000323 \times 10^{-7} t_i^2 I + 0.00005051 I U^2 - 0.00000313 t_o I U$$

(7.1)

Table 7.4 ANOVA table for the stepwise regression model of MRR

Source	DF	Sum of squares	Mean square	F value	$F_{0.001,7,17}$
Model	7	23719	3388.4741	61.37	6.22
Error	17	938.5955	55.2115		
Corrected Total	24	24657.5955			

7.5.2 Analysis of Taguchi method for the TWR

Table 7.5 presents the results of the ANOVA and F-test for the TWR. As shown in Table 7.5, open voltage, discharge current and pulse duration are the significant parameters on the TWR, whereas pulse interval is the less significant parameter on the TWR.

Table 7.5 Results of the ANOVA and F-test for the TWR

Parameter	Sum of squares	Degree	Mean square	F
Pulse duration	2072.245	4	518.0612	8.51**
Pulse interval	803.363	4	200.8406	3.30*
Discharge current	2514.073	4	628.5183	10.32**
Open voltage	3051.833	4	762.9582	12.53**
Error	487.184	8	60.8980	
Total	8928.698	24		

* Less significant parameter
** Significant parameter

The average values of the TWR at levels 1, 2, 3, 4, and 5 of the four input machining parameters are calculated to investigate the parameter effects. The effects of machining parameters on the average TWR are shown in Fig. 7.17. It can be seen from Fig. 7.17 that the TWR decreases with an increase in pulse duration, and it increases with an increase in pulse interval, discharge current, and open voltage, respectively. There are many reasons causing these phenomena. During EDM, a deposition layer can form on the electrode surface due to the decomposition of the dielectric and workpiece material attached to the tool electrode surface, and the tool wear can be prevented by the protective effects of the deposition layer [23,24]. As the pulse duration increases, the discharge energy delivered to the machining gap increases, the dielectric and workpiece are heated for more time, the released carbon decomposed from the dielectric is easily attached to the tool surface, the deposition effect is enhanced, the tool wear decreases; therefore, the TWR decreases. As the pulse interval increases, the time for deionization of the dielectric increases, the discharge energy delivered to the machining gap decreases in a unit time, the released carbon decomposed from the dielectric decreases, the deposition effect weakens, the tool wear compensation decreases; therefore, the TWR increases. In addition, the single pulse energy, thermal energy density and discharge explosive force increase with an increase in discharge current and open voltage, respectively, which enhances the tool material removal, so the TWR increases with an increase in discharge current and open voltage, respectively. The optimum machining parameters of the compound process for the TWR obtained from Fig. 7.17 are as follows: 400μs pulse duration; 300μs pulse interval; 15A discharge current; and 90V open voltage.

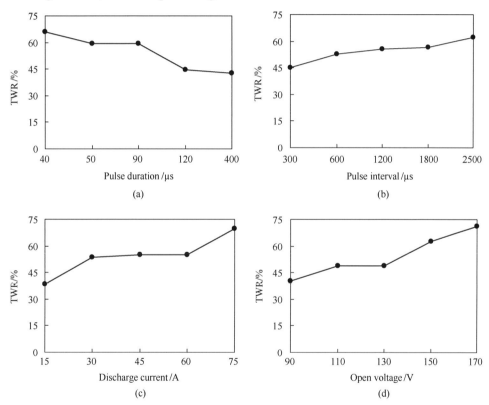

Fig. 7.17 Effects of machining parameters on the average TWR. (a) Effect of pulse duration on the average TWR, (b) Effect of pulse interval on the average TWR, (c) Effect of discharge current on the average TWR, (d) Effect of open voltage on the average TWR

The results of the analysis of variance for the stepwise model of the TWR are shown in Table 7.6. As in this case F is far greater than $F_{0.001,5,19}$, it is accepted that the mathematical model of the TWR by using stepwise regression analysis method is appropriate for a significance level, α, of 0.001. The mathematical relation between the TWR and machining parameters is as follows:

$$TWR = 34.38362 - 0.29323t_i - 0.00050015t_i^2 + 0.00005010t_o U + 0.00315 IU + 0.00068065 U^2 \quad (7.2)$$

Table 7.6 ANOVA table for the stepwise regression model of TWR

Source	DF	Sum of squares	Mean square	F value	$F_{0.001,5,19}$
Model	5	7457.4520	1491.4904	19.26	6.62
Error	19	1471.2457	73.4340		
Corrected Total	24	8928.6977			

7.5.3 Analysis of Taguchi method for the SR

The results of the ANOVA and F-test for the SR are given in Table 7.7. It can be seen from Table 7.7 that pulse duration, open voltage, and discharge current are the more significant parameters on the SR, whereas pulse interval is the significant parameter on the SR. When the pulse duration, open voltage, and discharge current are set at the high level, the huge discharge energy is delivered into the machining zone within a single pulse, and the discharge crater becomes larger and deeper. Therefore, the SR is high using the high pulse duration, open voltage, and discharge current, respectively.

Table 7.7 Results of ANOVA and F-test for the SR

Parameter	Sum of squares	Degree	Mean square	F
Pulse duration	21.976	4	5.4939	60.20***
Pulse interval	2.808	4	0.7020	7.69**
Discharge current	5.443	4	1.3607	14.91***
Open voltage	7.090	4	1.7726	19.42***
Error	0.730	8	0.0913	
Total	38.047	24		

** Significant parameter
*** More significant parameter

The average values of the SR at levels 1, 2, 3, 4, and 5 of the four input machining parameters are calculated to investigate the parameter effects. The effects of machining parameters on the average SR are shown in Fig. 7.18. The SR increases with an increase in pulse duration, discharge current, and open voltage, respectively, and it decreases with an increase in pulse interval. The phenomenon can be explained as follows. The single pulse energy increases with an increase in pulse duration, discharge current, and open voltage, respectively, the material removed by a single pulse increases with the single pulse energy increasing, the discharge crater becomes larger and deeper; therefore, the SR increases. In addition, with a longer pulse interval, there is more time to grind the modified surface layer on the workpiece and clear the disintegrated particles from the discharge gap, so the SR is low. The optimum machining parameters of the compound process for SR obtained from Fig. 7.18 are as follows: 40μs pulse duration; 2500μs pulse interval; 15A discharge current; and 90V open voltage.

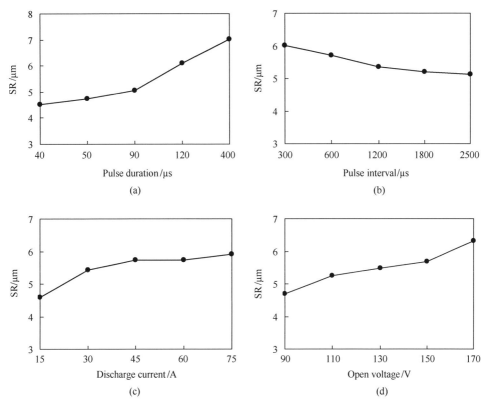

Fig. 7.18 Effects of machining parameters on the average SR. (a) Effect of pulse duration on the average SR, (b) Effect of pulse interval on the average SR, (c) Effect of discharge current on the average SR, (d) Effect of open voltage on the average SR

The results of the analysis of variance for the stepwise model of the SR are shown in Table 7.8. As in this case F is far greater than $F_{0.001,7,17}$, it is accepted that the mathematical model of SR by using stepwise regression analysis method is appropriate for a significance level, α, of 0.001. The mathematical relation between the SR and machining parameters is as follows:

$$SR=2.87450 - 0.00120t_o + 0.00001693t_o I + 0.00000234t_i U^2 - 2.62079 \times 10^{-9} t_i^2 U^2 - 5.72796 \times 10^{-9} t_i U^3$$

(7.3)

Table 7.8 ANOVA table for the stepwise regression model of the SR

Source	DF	Sum of squares	Mean square	F value	$F_{0.001,7,17}$
Model	7	35.7741	5.1106	38.22	6.22
Error	17	2.2729	0.1337		
Corrected Total	24	38.0470			

7.5.4 Confirmation experiments

The optimal combination level of machining parameters has been determined in the previous analysis. Table 7.9 shows the results of confirmation experiments for the compound process. As shown in this table, the experimental values agree well with predictions. Furthermore, the MRR, the TWR, and the SR can reach 138.3348mm³/min, 8.1797%, and 1.0543μm, respectively. Therefore, the experimental results confirm the optimization of the machining parameters and the improvement of the machining characteristics.

Table 7.9 Results of the confirmation experiments for the compound process

	Optimal machining parameters				Predicted	Experimental	Relative error/%
	t_i/μs	t_o/μs	I/A	U/V			
MRR/(mm³/min)	400	300	75	170	136.8814	138.3348	1.0506
TWR/%	400	300	15	90	8.2341	8.1797	0.6651
SR/μm	40	2500	15	90	1.0665	1.0543	1.1572

7.6 Surface integrity of SiC ceramic machined by the compound machining

7.6.1 Surface morphology of the machined surface

Surface morphology plays an important part in understanding the characteristic of the machined surface. The SEM micrographs of the machined surface at rough machining mode with different tool polarities are illustrated in Fig. 7.19. The micrographs reveal the appearance of many craters, droplets and cavities on the machined surface when rough machining mode is used. These phenomena point out that the SiC ceramic is molten and/or evaporated by the sparking thermal energy. During machining, sparks are formed at the electrical conductive phase on the SiC ceramic. The discharged energy produces very high temperatures at the point of the spark, causing a minute part of the workpiece to melt and vaporize. When the current flow ceases, a violent collapse of the plasma channel and vapor bubble causes superheated, molten liquids to explode into the gap and the dielectric liquid solidifies the molten material immediately. However, not all of the molten material can be removed because of the surface tension, tensile strength and bonding force between liquid and solid. The molten material remained on the workpiece surface is rapidly quenched by the emulsion and re-solidifies on the workpiece surface to form droplets. This resolidification material simultaneously shrinks after sparking due to the emulsion cooling. Regions where the molten material is re-solidified later on do not have enough molten material to fill in, which lead to cavities.

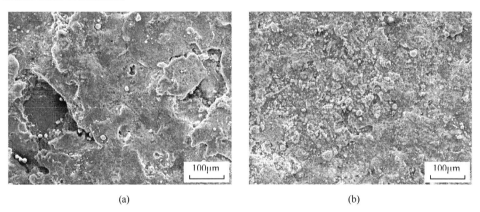

(a) (b)

Fig. 7.19 SEM micrographs of SiC ceramic surfaces machined by the compound process at rough machining mode with different tool polarities under the machining conditions: pulse duration of 400μs, pulse interval of 300μs, peak voltage of 150V, and peak current of 75A. (a) tool (–), (b) tool (+)

It is obvious that the craters and cavities are bigger and deeper in negative tool polarity, whereas the droplets are more in positive tool polarity under the same conditions. During the compound process, the tool rotates at a high speed, the discharge point transfer velocity between the electrode and workpiece is very high, and the continuance time of the discharge point between the certain point of the electrode and the certain point of the workpiece is very short. Because the mass of the electrons is much smaller than that of positive ions, and they can be accelerated quickly during a short time, the bombardment effect by electrons is stronger than that by positive ions; therefore, the craters and cavities on the workpiece surface produced by electrons are bigger and deeper. The bombardment effect by positive ions is weaker, so less molten material can be expelled during machining, and more droplets form on the workpiece surface.

The SEM micrograph of the machined surface at finish machining mode is shown in Fig. 7.20. Compared with Fig. 7.20(b), the machined surface is smooth at finish machining mode, and covered by fewer little craters and pockmarks. Moreover, it can be found that there are some grinding traces on the machined surface, which means that the material is mainly removed by mechanical grinding at finish machining mode.

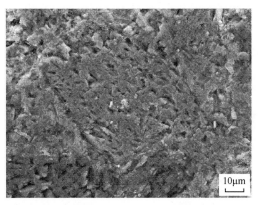

Fig. 7.20 SEM micrograph of SiC ceramic surface machined by the compound process at finish machining mode under the conditions: pulse duration of 40μs, pulse interval of 2500μs, peak voltage of 90V, peak current of 15A, and positive tool polarity

7.6.2 Surface roughness

Surface roughness is a critical parameter for evaluating the machined quality. To determine the effect of machining parameters on the SR of the SiC ceramic, the arithmetical mean deviation of the profile (R_a) is selected as the surface roughness parameter, and it is measured by a SR tester through averaging five measurements made stochastically at different positions on the machined surface.

Table 7.10 shows the experimental results of the SR with different machining conditions. From these results it is obvious that the negative tool polarity causes a poorer surface finish. As mentioned above, the bombardment effect by the electrons is stronger than that by positive ions during the compound machining of SiC ceramic. Stronger electrons bombardment results in bigger and deeper craters, leading to a rougher surface; therefore, the surface roughness is higher in negative tool polarity than that in positive tool polarity.

Table 7.10 also shows that the SR increases with the increase of pulse duration and peak current, respectively. This phenomenon can be explained by the correlation between the surface roughness and the machining parameters [Equation (4.3)]. Under a certain machining condition, the R_{max} increases with the increase of pulse duration and peak current, respectively; therefore, the SR increases.

Based on the surface roughness values in Table 7.10, it can be inferred that an excellent machined surface finish can be obtained by setting the machine parameters at a short pulse duration, a low peak current, and positive tool polarity.

Table 7.10 Process performance and surface composition with different machining conditions

Pulse duration/μs	Pulse interval/μs	Peak voltage/V	Peak current/A	Tool polarity	MRR /(mm³/min)	SR/μm	Fe/wt%	Cu/wt%
400	300	150	75	Negative	130.72	7.62	4.20	4.45
400	300	150	75	Positive	75.75	4.98	12.28	12.50
90	1200	110	45	Negative	33.21	4.28	1.46	1.03
90	1200	110	45	Positive	18.42	2.62	4.58	4.49
120	300	150	75	Negative	110.68	6.69	1.67	1.63
40	300	150	75	Negative	60.84	4.91	1.33	0.94
50	300	150	15	Negative	32.53	4.28	0.21	0.34
50	300	150	45	Negative	62.64	5.56	1.61	1.11

7.6.3 Micro-cracks on the machined surface

Fig. 7.21 and Fig. 7.22 show the micro-cracks on the SiC ceramic surfaces machined by the compound process under different pulse duration and peak current settings, respectively. During machining, the thermal impacts of the sparks are accompanied by a very rapid quench rate of the heated material in the plasma contact zone. These thermal waves cause expansion and contraction of re-solidified and heat affected material, which is the main reason for the appearance of the micro-cracks. The SiC ceramic has extreme hardness and brittleness, the gradient of thermal stress is very large during machining, so surface micro-cracks are easily generated on the machined surface. The micro-cracks on the machined surface lead to thermal spalling of the SiC ceramic, so the removal mechanism in the compound machining of SiC ceramic consists of not just the melting and evaporation, but also thermal spalling.

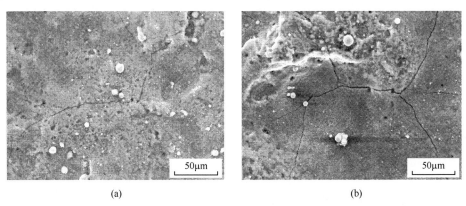

Fig. 7.21 SEM micrographs showing micro-cracks on the machined surfaces with different pulse durations under the machining conditions: Pulse interval of 300μs, peak voltage of 150V, peak current of 75A, and negative tool polarity. (a) Pulse duration of 120μs, (b) Pulse duration of 400μs

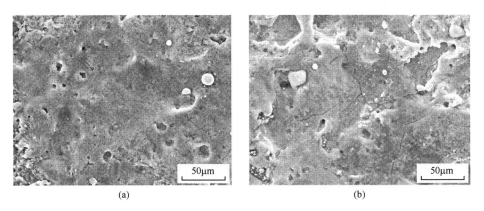

Fig. 7.22 SEM micrographs showing micro-cracks on the machined surfaces with different peak currents under the machining conditions: Pulse duration of 50μs, pulse interval of 300μs, peak voltage of 150V, and negative tool polarity. (a) Peak current of 15A, (b) Peak current of 45A

It can also be seen from Fig. 7.21 and Fig. 7.22 that the number and size of the micro-cracks on the machined surface increase with an increase in pulse duration and peak current, respectively. The phenomenon can be explained as follows: The amount of electrical discharge energy conducted into the machining gap increases with an increase in pulse duration and peak current, respectively, the energy density of the electrical discharge in the discharge spot on the machined surface increases, the thermal impact and thermal stress increase, so the number and size of the micro-cracks on the machined surface increase with an increase in pulse duration and peak current, respectively.

7.6.4 Compositions of the machined surface

EDS spectrum analysis is used to identify the elemental composition on the workpiece surface generated in different machining conditions. Fig. 7.23 and Fig. 7.24 show the EDS spectrum analysis of the machined surface and the unprocessed surface, respectively. The prominent elements on the machined surface are carbon (C), oxygen (O), silicon (Si), iron (Fe) and copper (Cu), whereas the prominent elements on the unprocessed surface are C, O and Si. This means that the migration of material from tool to workpiece occurs during the compound process.

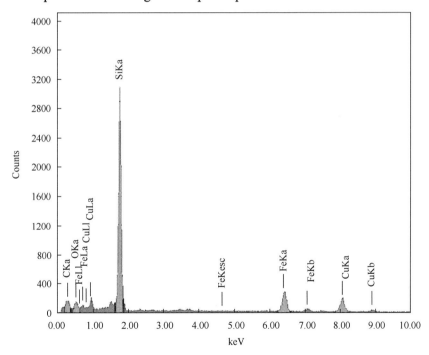

Fig. 7.23 EDS spectrum analysis of the machined SiC ceramic surface under the machining conditions: Pulse duration of 400μs, pulse interval of 300μs, peak voltage of 150V, peak current of 75A, and positive tool polarity

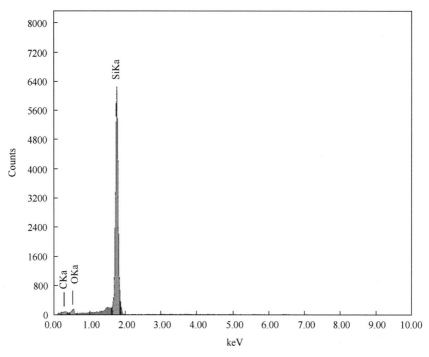

Fig. 7.24 EDS spectrum analysis of the unprocessed SiC ceramic surface

The Fe percentage and Cu percentage on the machined SiC ceramic surface with different machining conditions are presented in Table 7.3. Under the same conditions, the metal (Fe and Cu) percentage on the machined surface in positive tool polarity is higher than that in negative tool polarity. This phenomenon can be explained as follows: The electrodes and abrasive sticks contain Cu and Fe, the electrolysis reaction occurs easily in positive tool polarity under the application of the water-based emulsion, and the removed tool material is mostly ionized into metallic ion, which can attach to the workpiece surface easily. However, the electrolysis reaction occurs difficultly in negative tool polarity because the SiC is a nonmetal. Therefore, the metal percentage on the machined surface is high in positive tool polarity.

It can also be seen from Table 7.10 that the metal percentage on the machined surface increases with an increase in pulse duration and peak current, respectively. This is because the single pulse energy, thermal energy density and discharge explosive force increase with an increase in pulse duration and peak current, respectively, the tool material removal is enhanced, and more tool material can transfer to the workpiece surface; therefore, the metal percentage on the machined surface increases.

To study the phase change that has occurred during the compound process, the XRD patterns of the machined surface and the unprocessed surface are presented in Fig. 7.25 and Fig. 7.26, respectively. The prominent substances on the machined surface are Cu_3Si, $FeSi_2$, Si and SiC, whereas the prominent substances on the unprocessed surface are Si and SiC. This indicates that Cu_3Si and $FeSi_2$ are generated during the compound machining of SiC ceramic. It can be concluded that a modified layer forms on the machined surface, and a combination reaction takes place during the compound machining of SiC ceramic, which can be described as follows:

$$3Cu+Si \rightarrow Cu_3Si$$

$$Fe+2Si \rightarrow FeSi_2$$

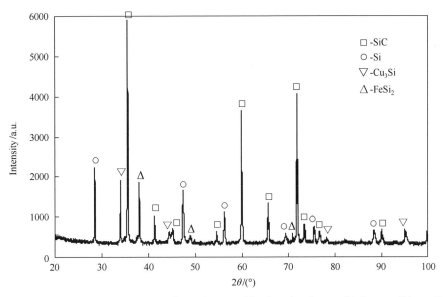

Fig. 7.25 XRD patterns of SiC ceramic surface machined under the machining conditions: Pulse duration of 400μs, pulse interval of 300μs, peak voltage of 150V, peak current of 75A, and positive tool polarity

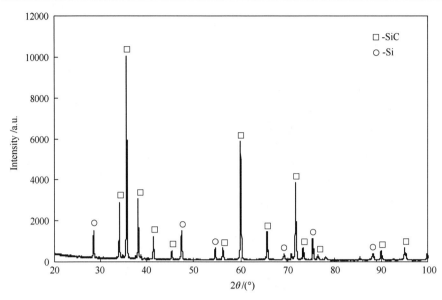

Fig. 7.26 XRD patterns of unprocessed SiC ceramic surface

7.7 Conclusions

End electrical discharge milling and mechanical grinding happen alternately and are mutually beneficial, so the compound process is able to effectively machine a large surface area of SiC ceramic with a good surface quality. The process also shows very good working environmental practice.

Higher MRR, lower TWR, and higher SR can be obtained with negative tool polarity and long pulse duration; the MRR, TWR, and SR increase with increasing open-circuit voltage and discharge current; the MRR, TWR, and SR decrease with increasing diamond grit size; and the MRR and TWR increase, but the SR decreases with increasing emulsion flux. Overall, the pulse interval of 300μs, emulsion concentration of 12wt%, milling depth of 0.1mm, and 10 tool sticks were suitable for the compound machining of SiC ceramic.

ANOVA and F-test of experimental data reveal that open voltage is the significant parameter on MRR, whereas discharge current and pulse interval are the less significant parameters on MRR. Moreover, machining parameter combination levels that maximized MRR are as follows: 400μs pulse duration (t_i); 300μs pulse interval (t_o); 75A discharge current (I); and 170V open voltage (U).

Experimental data analysis based on Taguchi method indicate that open voltage, discharge current and pulse duration are the significant parameters on the TWR, whereas pulse interval is the less significant parameter on the TWR. Moreover, machining parameter combination levels that minimized the TWR are as follows: 400μs pulse duration; 300μs pulse interval; 15A discharge current; and 90V open voltage.

Experimental data analysis based on Taguchi method demonstrate that pulse duration, open voltage, and discharge current are the more significant parameters on SR, whereas pulse interval is the significant parameter on the SR. Moreover, machining parameter combination levels that optimized SR are as follows: 40μs pulse duration; 2500μs pulse interval; 15A discharge current; and 90V open voltage.

With each optimal combination level of machining parameters, the MRR, TWR and SR can reach 138.3348mm^3/min, 8.1797% and 1.0543μm, respectively. The experimental values agree well with predictions, which confirm the validity of the used Taguchi method for enhancing the machining performance and optimizing the machining parameters.

When rough machining mode is used, the machined surface is rough, and the craters and cavities are bigger and deeper in negative tool polarity, whereas the droplets are more in positive tool polarity under the same conditions on the machined surface. When finish machining mode is used, the machined surface is smooth, and covered by fewer little craters and pockmarks. Moreover, SiC ceramic is mainly removed by end ED milling such as melting, evaporation and thermal spalling at rough machining mode, whereas it is mainly removed by mechanical grinding at finish machining mode.

As the pulse duration and peak current increase, respectively, the surface roughness and the number and size of the micro-cracks on the machined surface increase. Furthermore, the surface roughness is higher in negative tool polarity than that in positive tool polarity. Therefore, the positive tool polarity, the short pulse duration, and the low peak current cause a fine surface finish.

The chemical compositions of the machined surface differ from the unprocessed surface due to the diffusion of the tool material, and a combination reaction takes place during machining. Furthermore, the tool material that transfers to the workpiece surface increases with the increase of pulse duration and peak current, respectively, and more tool material can transfer to the workpiece surface in case of positive tool polarity, when compared to negative tool polarity under the same conditions.

References

[1] Guo X Z, Yang H, Zhang L J, et al. Sintering behavior, microstructure and mechanical properties of silicon carbide ceramics containing different nano-TiN additive. Ceramics International, 2010, 36: 161-165.

[2] Bae H T, Choi H J, Jeong J H, et al. The effect of reaction temperature on the tribological behavior of the surface modified silicon carbide by the carbide derived carbon process. Materials and Manufacturing Processes, 2010, 25: 345-349.

[3] Yu X M, Zhou W C, Luo F, et al. Effect of fabrication atmosphere on dielectric properties of SiC/SiC composites. Journal of Alloys and Compounds, 2009, 479: L1-L3.

[4] Okada A. Automotive and industrial applications of structural ceramics in Japan. Journal of the European Ceramic Society, 2008, 28: 1097-1104.

[5] Krnel K, Stadler Z, Kosmac T. Carbon/carbon-silicon-carbide dual-matrix composites for brake discs. Materials and Manufacturing Processes, 2008, 23: 587-590.

[6] Katoh Y, Kondo S, Snead L L. DC electrical conductivity of silicon carbide ceramics and composites for flow channel insert applications. Journal of Nuclear Materials, 2009, 386-388: 639-642.

[7] Chen J Y, Shen J Y, Huang H, et al. Grinding characteristics in high speed grinding of engineering ceramics with brazed diamond wheels. Journal of Materials Processing Technology, 2010, 210: 899-906.

[8] Yan J, Zhang Z, Kuriyagawa T. Mechanism for material removal in diamond turning of reaction-bonded silicon carbide. International Journal of Machine Tools and Manufacture, 2009, 49: 366-374.

[9] Chen Y F, Lin Y J, Lin Y C, et al. Optimization of electrodischarge machining parameters on ZrO_2 ceramic using the Taguchi method. Proceedings of the Institution of Mechanical Engineers, Part B: Journal of Engineering Manufacture, 2010, 224: 195-205.

[10] Marafona J D, Araujo A. Influence of workpiece hardness on EDM performance. International Journal of Machine Tools and Manufacture, 2009, 49: 744-748.

[11] Govindan P, Joshi S S. Experimental characterization of material removal in dry electrical discharge drilling. International Journal of Machine Tools and Manufacture, 2010, 50: 431-443.

[12] Kato T, Noro T, Takahashi H, et al. Characterization of electric discharge machining for silicon carbide single crystal. Materials Science Forum, 2009, 600-603: 855-858.

[13] Luis C J, Puertas I, Villa G. Material removal rate and electrode wear study on the EDM of silicon carbide. Journal of Materials Processing Technology, 2005, 164/165: 889-896.

[14] Shih H R, Shu K M. A study of electrical discharge grinding using a rotary disk electrode. International Journal of Advanced Manufacturing Technology, 2008, 38: 59-67.

[15] Shu K M, Tu G C. Study of electrical discharge grinding using metal matrix composite electrodes. International Journal of Machine Tools and Manufacture, 2003, 43: 845-854.

[16] Liu Y H, Ji R J, Li Q Y, et al. An experimental investigation for electric discharge milling of SiC ceramics with high electrical resistivity. Journal of Alloys and Compounds, 2009, 472: 406-410.

[17] Lauwers B, Kruth J P, Brans K. Development of technology and strategies for the machining of ceramic components by sinking and milling EDM. CIRP Annals - Manufacturing Technology, 2007, 56: 225-228.

[18] Han F Z, Wang Y X, Zhou M. High-speed EDM milling with moving electric arcs. International Journal of Machine Tools and Manufacture, 2009, 49: 20-24.

[19] Ji R J, Liu Y H, Yu L L, et al. Study on high efficient electric discharge milling of silicon carbide ceramic with high resistivity. Chinese Science Bulletin, 2008, 53: 3247-3254.

[20] Petrofes N F, Gadalla A M. Electrical discharge machining of advanced ceramics. American Ceramic Society Bulletin, 1988, 67: 1048-1052.

[21] Mei C L, Fan J C. Data analysis method. Beijing: Higher Education Press, 2006: 267-281.

[22] Kunieda M, Kobayashi T. Clarifying mechanism of determining tool electrode wear ratio in EDM using spectroscopic measurement of vapor density. Journal of Materials Processing Technology, 2004, 149: 284-288.

[23] Marafona J. Black layer characterisation and electrode wear ratio in electrical discharge machining (EDM). Journal of Materials Processing Technology, 2007, 184(1-3): 27-31.

[24] Abdulkareem S, Ali K A, Konneh M. Cooling effect on electrode and process parameters in EDM. Materials and Manufacturing Processes, 2010, 25(6): 462-466.

Chapter 8 Machining Fluid for Electrical Discharge Machining of Engineering Ceramics

Machining fluid is a primary factor that affects the MRR, surface quality, and electrode wear of EDM. Kerosene is the most commonly used as the working fluid in EDM, but it shows low ignition temperature and high volatility, if the improper operations are undertaken, conflagration can be caused. Using distilled water or pure water as the machining fluid in EDM, without fire hazard occurs, the working environmental is well, however, using distilled water or pure water as the machining fluid in EDM, the material removal rate of machining large surface is low, the machine tool is easy to be eroded.

Three kinds of emulsion are proposed as the dielectric during electrical discharge machining of engineering ceramics in this chapter. The effects of dielectric, tool polarity, pulse duration, pulse interval, peak voltage, and peak current on the process performance such as the MRR and SR have been investigated. Furthermore, the microstructure of the machined surface is examined with SEM, EDS and XRD.

8.1 Introduction

In EDM, the working fluid plays an important role affecting the MRR and the properties of the machined surface. The essential tasks of the working fluid are transporting removal particles, increasing energy density of plasma channel, and cooling electrodes etc [1]. Most die-sinking EDM processes use kerosene as the EDM dielectric fluid, but kerosene-relevant properties are degraded during long-term machining. Another disadvantage of using kerosene is that kerosene pollutes the air and the high discharge temperature decomposes the kerosene, which causes carbon elements to adhere on the electrode surface. The adhered carbon elements affect normal discharge [2,3]. The powders such as aluminum (Al), Si, alumina and nickel (Ni) are added into kerosene for EDM, the low surface roughness and high corrosion resistance of the workpiece can be obtained [4-6]. In comparison with kerosene, water used as machining fluid for EDM shows low cost, fine heat conductivity and no pollution [7]. Chen et al. [8] employed distilled water to machine Ti-6Al-4V alloy in EDM, the high material removal rate and the low REWR have been obtained. Chung et al. [9] discovered that using deionized water instead of kerosene could reduce the electrode wear and increase the machining rate in EDM of micro hemisphere. Chow et al. [10,11] added SiC powder into pure water as the working fluid for micro-slit EDM. The results indicated that the addition of SiC powder would increase electrical conductivity of the working fluid, enlarge the discharge gap between electrode and workpiece, and also extrude debris easily, therefore, which would increase the material removal rate. However, using distilled water or pure water as the machining fluid in EDM the material removal rate of machining large surface is low, the machine tool is easy to be eroded.

Emulsion has been widely used as the working fluid in WEDM. However, it is very little used in die sinking EDM. This is because that emulsion used in die sinking EDM shows low efficiency.

The authors had developed a process of machining SiC ceramic using ED milling [12, 13]. ED milling used a steel toothed wheel as the tool electrode, and employed the pulse generator used in EDM. It is able to effectively machine a large surface area on SiC ceramic. The dielectric plays an important role in ED milling. This chapter proposes three kinds of emulsion as the dielectric. The effects of composition and concentration of emulsifier on the emulsion property and EDM performance have been investigated. The effects of dielectric, tool polarity, pulse duration, pulse interval, peak voltage, and peak current on the process performance such as the MRR and SR have been investigated.

8.2 Experimental procedures

In this chapter, the emulsifiers and the machine oil are heated in an electrical constant temperature water bath pot (HH-S4, Jintan Medical Instrument Factory, China) and were emulsified with an ultrasonic processor (FS-250, Shanghai Shengxi Ultrasonic Instrument Corporation, China), then the emulsified oil is created. The emulsion used in the following experiments is composed of 5wt% emulsified oil and 95wt% distilled water, which are mixed with a constant speed power-driver mixer (JJ-2, Jintan Medical Instrument Factory, China). The electrical conductivity of the emulsion is measured by a conductivity meter (HI8733, Hanna, Italy). The surface tension of the emulsion is measured by a surface tensiometer (Jzhy1-180, Chengteh Mingwei Testing Machine Ltd, China).

The MRR and electrode wear are obtained through measuring the mass of the workpiece and the electrode before and after machining with an electron balance (BS224S, Sartorius, Germany). The SR is measured by a surface roughness tester (2205A, Harbin Measuring & Cutting Tool Group Co. Ltd., China). The microstructure of the workpiece surface is observed by SEM (SEM FEI.QUANTA 200, Holand). The discharge gap is measured by a dial indicator (803-01, Harbin Measuring & Cutting Tool Group Co. Ltd., China). At first scale value of the dial indicator (h_0) is obtained as the electrode tool contacts the workpiece without pulse voltage. Then the electrode tool is moved up 1mm with the machine tool spindle. Last the pulse generator connecting the electrode tool and workpiece is turned on, the electrode tool is moved down with the machine tool spindle, the scale value of the dial indicator (h_1) is obtained at the moment that the electrical discharge is produced. The discharge gap equals to h_1 minus h_0.

8.3 Effect of the additive on the emulsion property and EDM performance

8.3.1 Effect of the anionic compound emulsifier on the emulsion property

The anionic compound emulsifier (ACE) is made of sodium petroleum sulfonate and triethanolamine oleate. The dosages of sodium petroleum sulfonate and triethanolamine oleate are determined by the hydrophile-lipophile balance value of emulsifier and machine oil. The effect of ACE concentration in the emulsified oil on electrical conductivity (EC) of the emulsion is shown in Fig. 8.1. The figure indicates that EC of the emulsion increases with an increase in ACE. The reason for this is that the ACE is water-soluble compound, it can ionize when it is dissolved into water, the ion concentration in the emulsion increase with increasing ACE; therefore, the EC increases.

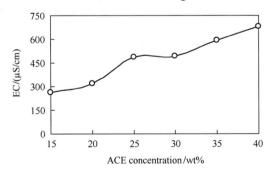

Fig. 8.1 Effect of ACE concentration on EC

Fig. 8.2 shows the relationship between surface tension of the emulsion and ACE concentration in the emulsified oil. The figure indicates that the surface tension initially decreases rapidly with an increase of ACE, and then decreases very slowly with an increase in ACE. The reason for this is that the molecules of ACE adsorb on the emulsion surface, the hydrophilic groups are dissolved in water and the hydrophobic groups are dispersed on the emulsion surface. The van der Waals force between the hydrophobic groups is small [14]. When less surfactant is added, the adsorption effect increases and the van der Waals force decreases with an increase in ACE, so the surface tension decreases rapidly with increasing ACE. When the ACE concentration exceeds 25wt%, the surfactant achieves critical micelle concentration, the adsorption of the surfactant molecules on the emulsion surface is saturated; therefore, the surface tension decreases very slowly.

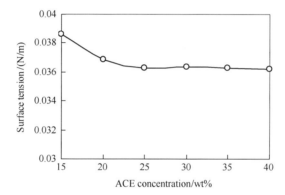

Fig. 8.2 Effect of ACE concentration on surface tension

8.3.2 Effect of ACE concentration on EDM performance

The effect of ACE concentration in the emulsified oil on the MRR is shown in Fig. 8.3. The MRR initially increases with an increase in ACE and then decreases with increasing ACE. There are many reasons causing this phenomenon. As the ACE concentration is lower than 25wt%, the surface tension decreases rapidly with an increase of ACE, permeability of the emulsion increases with increasing the surface tension, the emulsion enters into the discharge gap easily, the EDM process becomes stability; therefore, the MRR is high. As the ACE concentration is higher than 25wt%, the surface tension decreases very slowly, permeability of the emulsion increases very slowly, but the EC of the emulsion is high, the discharge gap is large, the breakdown voltage decreases largely; therefore, the MRR is low.

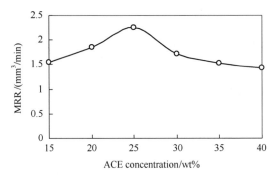

Fig. 8.3　Effect of ACE concentration on MRR

Fig. 8.4 shows the effect of ACE on the SR. The SR decreases with increasing ACE. There are many reasons causing this phenomenon. ACE can be ionized when it is dissolved in water, ion concentration in the emulsion increases with increasing ACE, the discharge gap becomes large, the breakdown voltage decreases, single input pulse generates several discharge spots because the ion concentration in the emulsion increases, the energy density of the discharge channel decreases, the crater generated by a single pulse is shallow; therefore, the SR decreases with an increase in ACE concentration.

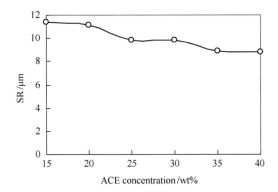

Fig. 8.4　Effect of ACE concentration on SR

Fig. 8.5 shows the effect of ACE on wear ratio of the tool electrode. The wear ratio is calculated as the ratio of the weight of material removed from the electrode to the weight of material removed from the workpiece. It can be seen from Fig. 8.5 that the wear ratio decreases with increasing ACE. There are many reasons causing this phenomenon. The permeability of the emulsion and the discharge gap increase with increasing ACE, the machining fluid easily enter into the discharge gap, the cool effect of the machining fluid on the tool electrode increases, the eroded materials are easy to be washed away, the EDM process becomes stability; therefore, the wear ratio decreases.

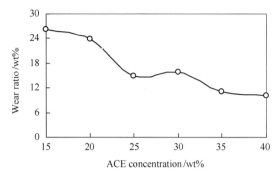

Fig. 8.5　Effect of ACE concentration on wear ratio

Fig. 8.6 shows the relationship between discharge gap and ACE concentration in the emulsified oil. The discharge gap increases with an increase in ACE concentration. This is because the electrical conductivity of the emulsion increases with increasing ACE, the breakdown voltage of the emulsion decreases, so the discharge gap increases.

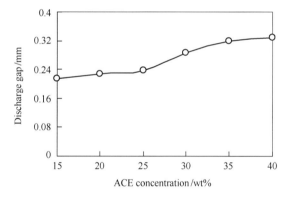

Fig. 8.6　Effect of ACE concentration on discharge gap

8.3.3　Effect of OP-10 concentration on emulsion property

The washing capability of emulsion has a great effect on EDM. In order to improve the washing capability of emulsion, the OP-10 is added into the developed emulsion.

The effect of OP-10 concentration in the emulsified oil with 25wt% ACE on EC of the emulsion is shown in Fig. 8.7. The EC of the emulsion changes very little with an increase in OP-10. This is because OP-10 is a nonionic surfactant; it can't ionize when it is dissolved in water [14]. The ion in the emulsion does not change with increasing OP-10 concentration, so the EC changes very little.

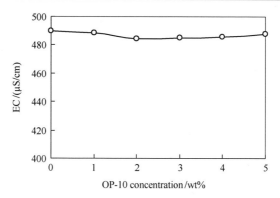

Fig. 8.7 Effect of OP-10 concentration on electrical conductivity

Fig. 8.8 shows that surface tension of emulsion decreases very slowly with increasing OP-10 concentration in the emulsified oil with 25wt% ACE. The reason for this is that OP-10 molecule can insert into the saturated adsorption layer formed by ACE, it can improve the adsorption action and decrease the van der Waals force between the hydrophobic groups, so the surface tension decreases very slowly with increasing OP-10 concentration.

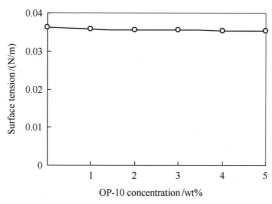

Fig. 8.8 Effect of OP-10 concentration on surface tension

8.3.4 Effect of OP-10 concentration on EDM performance

The effect of OP-10 concentration in the emulsified oil with 25wt% ACE on the MRR is shown in Fig. 8.9. It can be seen from this figure; the MRR initially increases with an increase in OP-10 and then decreases with increasing OP-10. There are many reasons causing the phenomena. OP-10 has a good washing capability, the eroded materials by EDM can be washed away easily, and the electrical discharge becomes stability; so the MRR increases with increasing OP-10. The emulsion with OP-10 concentration more than 1wt% shows high foaming ability, air bubbles are easily generated by EDM in the emulsion. It makes the viscosity and pinch effect of the emulsion decrease. The density of discharge energy decreases; therefore, the MRR is low.

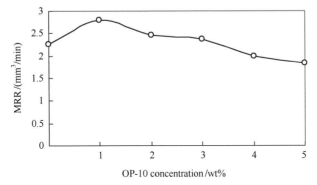

Fig. 8.9　Effect of OP-10 concentration on the MRR

As shown in Fig. 8.10, SR decreases slowly with increasing OP-10 concentration in the emulsified oil with 25wt% ACE. The reason for this is that OP-10 has a good washing capability, the eroded materials can be washed away easily, the electrical discharge becomes stability; therefore, the SR is low.

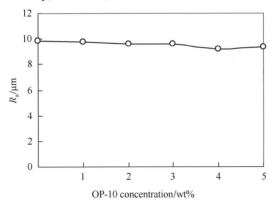

Fig. 8.10　Effect of OP-10 concentration on SR

Fig. 8.11 shows the effect of OP-10 concentration in the emulsified oil with 25wt% ACE on wear ratio of the tool electrode. The wear ratio increases with an increase in OP-10 concentration. This is because the good washing capability of the emulsion makes the protective film on the electrode surface be generated by EDM difficultly, the electrode wear increases; therefore, the wear ratio increases with increasing OP-10 concentration.

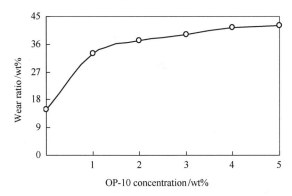

Fig. 8.11 Effect of OP-10 concentration on wear ratio

As shown in Fig. 8.12, discharge gap initially decreases with an increase of OP-10 concentration in the emulsified oil with 25wt% ACE and then increases slowly with increasing OP-10 concentration. There are many reasons causing this phenomenon. With a less OP-10 concentration, the machining fluid can be emulsified fully; the insulation intensity of the machining fluid is high; therefore, the discharge gap decreases with an increase in OP-10 concentration. It can be seen from Fig. 8.12; as the OP-10 concentration is higher than 2wt% EC of the emulsion increases slowly with an increase in OP-10 concentration; therefore, the discharge gap increases slowly with increasing OP-10 concentration.

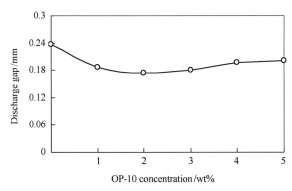

Fig. 8.12 Effect of OP-10 concentration on discharge gap

8.4 Influence of dielectric on the process performance for ED milling of SiC ceramic

In the emulsion, the ACE made of sodium petroleum sulfonate and triethanolamine oleate is used as the emulsifier, OP-10 is used as the cleaner, and the rosin is used as the blasting agent. Emulsion-1 is 95wt% distilled water+5wt% emulsified oil with 25wt% ACE and 75wt% machine oil. Emulsion-2 is 95wt% distilled water+5wt% emulsified oil with 25wt% ACE, 1wt% OP-10, and 74wt% machine oil. Emulsion-3 was 95wt% distilled water+5wt% emulsified oil with 25wt% ACE, 1wt% OP-10, 2wt% rosin, and 72wt% machine oil.

The microstructure of the silicon carbide ceramic matrix is illustrated in Fig. 8.13 with a metallographic microscope (Nikon-300, Nikon Corporation, Japan). In this image the gray grains are SiC. The white dots spread around the SiC grains are free Si. Among the silicon, the smaller gray dots are also SiC generated from the reaction of Si and C during the sintering process. The black granules are residual C [15]. Table 8.1 presents the physical and mechanical properties of the SiC ceramic in this experiment.

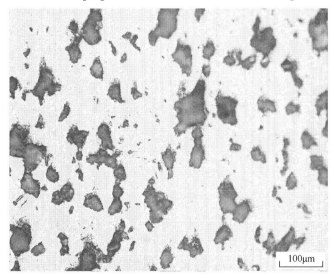

Fig. 8.13 Microstructure of the silicon carbide ceramic

Table 8.1 Physical and mechanical properties of silicon carbide cerami

Item	Data
Maximum service temperature/℃	1380
Density/(g/cm^3)	3.02
Open porosity/%	<0.1
Bending strength/MPa	250 (20℃)
	280 (1200℃)

Continued

Item	Data
Elastic modulus/GPa	330 (20℃)
	300 (1200℃)
Thermal conductivity/(W/(m · K))	45 (1200℃)
Thermal expansion coefficient/($10^{-6}K^{-1}$)	4.5
Mohs hardness	13
Acid and alkali resistance	Excellent

8.4.1 Effect of tool polarity on the process performance in different emulsions

Tool polarity is a primary factor that affects the process performance. The effects of tool polarity on the MRR and the SR with different emulsions are given in Fig. 8.14 and Fig. 8.15, respectively, for pulse duration of 50μs, pulse interval of 50μs, peak voltage of 200V, and peak current of 25A.

The MRR with different tool polarities and emulsions is given in Fig. 8.14. It can be seen from this figure that under the same conditions the MRR in positive tool polarity is higher than that in negative tool polarity. The phenomenon can be explained as follows. The positive ions in the discharge channel have enough time to be accelerated with a long pulse duration. Because the mass of the positive ions is much larger than that of the electrons, the bombardment effect by the positive ions is stronger than that by the electrons; therefore, the MRR is high in positive tool polarity. It can also be seen from Fig. 8.14 that under the same conditions the MRR with emulsion-3 is the highest, and the MRR with emulsion-1 is the lowest. There are many reasons causing this phenomenon. OP-10 has the good washing capability, the eroded materials can be washed away easily, and the electrical discharge becomes stable during ED milling; so the MRRs with emulsion-2 and emulsion-3 are higher than that with emulsion-1. Furthermore, the rosin is easily oxidated and inflammable. During ED milling, emulsion-3 containing the rosin will be locally gasified and the explosive mix gases are produced, which can cause the explosion, advance the density of discharge energy, improve the discharge force and enhance the removal of the eroded materials, so the MRR with emulsion-3 is higher than that with emulsion-2.

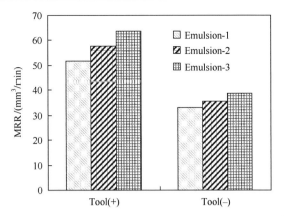

Fig. 8.14 Effect of tool polarity on the MRR with different emulsions

The effect of tool polarity and dielectric on the SR is shown in Fig. 8.15. Under the same conditions, the SR in negative tool polarity is higher than that in positive tool polarity. The phenomenon can be explained as follows. The SiC ceramic is composed of SiC particles distributing in the network of Si, the basis is SiC and the SiC chemical bond is stable. As mentioned above, the positive ions in the discharge channel have enough time to be accelerated during machining. The mass and energy of the positive ions are large, the SiC chemical bond can be easily ruptured by the positive ions, hence, the electrical discharges are stable, and the surface quality is good in positive tool polarity. Whereas the mass and energy of the electrons are small, the SiC chemical bond can't be easily destroyed by the electrons, the stability of the electrical discharges becomes bad, and arcs are easily generated in negative tool polarity; therefore, the SR is high in negative tool polarity. It can also be seen from Fig. 8.15 that under the same conditions the SR with emulsion-1 is the highest and the SR with emulsion-3 is the lowest. As mentioned above, emulsion-2 and emulsion-3 containing OP-10 have the good washing capability, the eroded materials are washed away easily, and the electrical discharge becomes stable during machining; so the SRs with emulsion-2 and emulsion-3 are lower than that with emulsion-1. In addition, the rosin is a weak acid, and it can ionize when it dissolves in the emulsion. The electrical conductivity and the acidity of the emulsion increase with an increase in rosin concentration, the electrolysis reaction will happen in emulsion-3 during ED milling, and the crater generated by a single pulse is shallow; therefore, the SR with emulsion-3 is lower than that with emulsion-2. The original contour curves of the SiC surface with different tool polarities and emulsions are shown in Fig. 8.16, in which the X-coordinate is the sampling length and the Y-coordinate is the contour depth. Because of those, positive tool polarity and emulsion-3 are suitable for ED milling of the SiC ceramic.

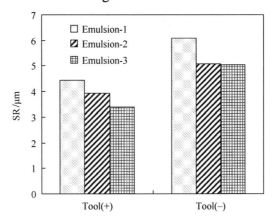

Fig. 8.15 Effect of tool polarity on SR with different emulsions

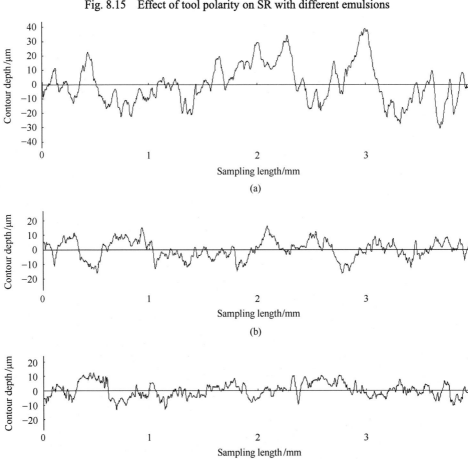

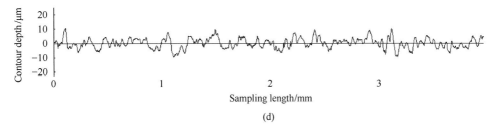

Fig. 8.16 Original contour curves of the machined silicon carbide surface with different tool polarities and emulsions. (a) Tool (−) emulsion-1, (b) Tool(+) emulsion-1, (c) Tool (+) emulsion-2, (d) Tool(+) emulsion-3

8.4.2 Effect of pulse duration on the process performance in different emulsions

The effects of pulse duration on the MRR and the SR in different emulsions are illustrated in Fig. 8.17 and Fig. 8.18, respectively, for pulse interval of 50μs, peak voltage of 200V, peak current of 25A, and positive tool polarity.

As shown in Fig. 8.17, the MRR with emulsion-3 is the highest and the MRR with emulsion-1 is the lowest under the same pulse duration. In addition, the MRR initially increases with an increase in pulse duration and then decreases with an increase in pulse duration. There are many reasons causing this phenomenon. As the pulse duration is short, the single pulse energy and the material removed by a single pulse increase with an increase in pulse duration; therefore, the MRR increases. However, when the pulse duration is longer than 50μs, the longer the pulse duration, the more thermal energy is lost due to the SiC ceramic heat conduction; therefore, the MRR decreases with an increase in pulse duration.

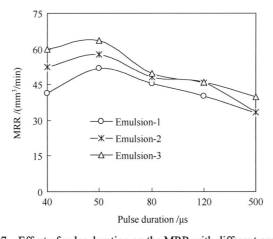

Fig. 8.17 Effect of pulse duration on the MRR with different emulsions

Fig. 8.18 shows the effect of emulsion and pulse duration on the SR. The SR with emulsion-1 is the highest and the SR with emulsion-3 is the lowest under the same pulse duration. Furthermore, the SR increases with an increase in pulse duration. This is because the crater size generated by a single pulse becomes larger with an increase in single pulse energy. The single pulse energy increases with an increase in pulse duration; therefore, the SR increases with an increase in pulse duration.

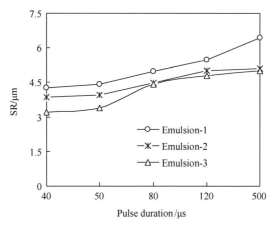

Fig. 8.18 Effect of pulse duration on the SR with different emulsions

8.4.3 Effect of pulse interval on the process performance in different emulsions

The effects of emulsion and pulse interval on the MRR and the SR in different emulsions are illustrated in Fig. 8.19 and Fig. 8.20, respectively, for pulse duration of 50μs, peak voltage of 200V, peak current of 25A, and positive tool polarity.

Fig. 8.19 shows the effect of dielectric and pulse interval on the MRR. The MRR with emulsion-3 is the highest and the MRR with emulsion-1 is the lowest under the same pulse interval. In addition, the MRR decreases gradually with an increase in pulse interval. This is because the pulse frequency decreases with an increase in pulse interval under a certain pulse duration. The discharge energy and the material removed in a unit time decrease with an increase in pulse interval, so the MRR decreases.

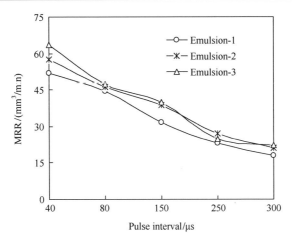

Fig. 8.19 Effect of pulse interval on the MRR with different emulsions

As shown in Fig. 8.20, the SR with emulsion-1 is the highest and the SR with emulsion-3 is the lowest under the same pulse interval. Furthermore, the SR initially increases with the increase of pulse interval and then decreases with the increase of pulse interval. This phenomenon can be explained as follows: As the pulse interval is short, a longer pulse interval means more time for deionization of the dielectric, the breakdown voltage and the discharge explosive force increase, the crater size generated by a single pulse becomes larger and deeper; therefore, the SR increases with an increase in pulse interval. However, when the pulse interval is longer than 80μs, the thermal energy is distributed on the material, the breakdown voltage and the discharge explosive force do not increase, the amount of crater generated by electrical discharges decreases; therefore, the SR decreases with an increase in pulse interval.

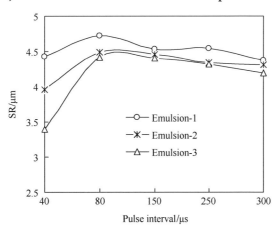

Fig. 8.20 Effect of pulse interval on the SR with different emulsions

8.4.4 Effect of peak voltage on the process performance in different emulsions

The effects of dielectric and peak voltage on the MRR and the SR in different emulsions are shown in Fig. 8.21 and Fig. 8.22, respectively, for pulse duration of 50μs, pulse interval of 50μs, peak current of 25A, and positive tool polarity.

Fig. 8.21 shows the effect of dielectric and peak voltage on the MRR. The MRR with emulsion-3 is the highest and the MRR with emulsion-1 is the lowest under the same peak voltage. In addition, the MRR increases with an increase in peak voltage. It is considered that the discharge current increases and the breakdown delay time decreases with an increase in peak voltage, so the single pulse energy increases. The material removed by a single pulse increases with an increase in single pulse energy; therefore, the MRR increases.

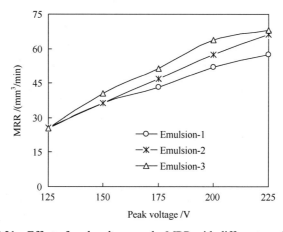

Fig. 8.21 Effect of peak voltage on the MRR with different emulsions

Fig. 8.22 shows the effect of emulsion and peak voltage on SR. The SR with emulsion-1 is the highest and the SR with emulsion-3 is the lowest under the same peak voltage. Furthermore, the SR increases slowly with an increase in peak voltage when it is less than 200V, and the SR increases rapidly when the peak voltage increases from 200V to 225V. There are many reasons causing the phenomena. As the peak voltage is lower than 200V, the crater size generated by a single pulse is small and shallow due to the high thermal decomposition point of the SiC ceramic; therefore, the SR increases slowly with an increase in peak voltage. However, when the peak voltage is higher than 200V, the discharge current becomes larger and the thermal energy density increases rapidly, the crater size generated by a single pulse becomes larger and deeper, and the material is even removed by explosion or flaking; therefore, the SR increases rapidly.

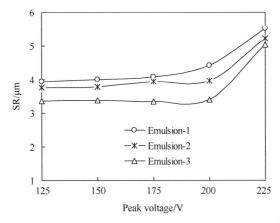

Fig. 8.22 Effect of peak voltage on the SR with different emulsions

8.4.5 Effect of peak current on the process performance in different emulsions

The effects of dielectric and peak current on the MRR and the SR in different emulsions are shown in Fig. 8.23 and Fig. 8.24, respectively, for pulse duration of 50μs, pulse interval of 50μs, peak voltage of 200V, and positive tool polarity.

As shown in Fig. 8.23, the MRR with emulsion-3 is the highest and the MRR with emulsion-1 is the lowest under the same peak current. In addition, the MRR increases gradually with an increase in peak current. This is because the single pulse energy and the material removed by a single pulse increase with an increase in peak current.

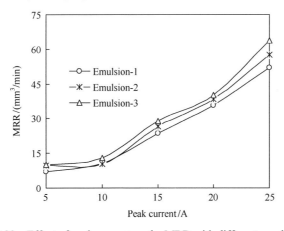

Fig. 8.23 Effect of peak current on the MRR with different emulsions

Fig. 8.24 shows the effect of dielectric and peak current on SR. The SR with emulsion-1 is the highest and the SR with emulsion-3 is the lowest under the same peak current. Furthermore, the SR increases gradually with the increase of peak current. The reason is that the single pulse energy and the material removed by a single pulse increase with an increase in peak current, the crater size generated by a single pulse becomes larger and deeper, so the SR increases with an increase in peak current.

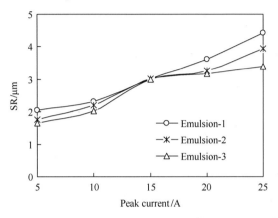

Fig. 8.24 Effect of peak current on the SR with different emulsions

8.4.6 Surface analysis of workpiece

1. Surface topography of the machined workpiece

The SEM micrographs of the SiC ceramic surface machined by ED milling with emulsion-1, emulsion-2 and emulsion-3 are shown in Fig. 8.25(a)~Fig. 8.25(c), respectively, for pulse duration of 50μs, pulse interval of 50μs, peak voltage of 200V, peak current of 25A, and positive tool polarity. It can be seen from Fig. 8.25(a) that there are many craters and grains left on the SiC ceramic surface machined by ED milling with emulsion-1. As shown in Fig. 8.25(b), there are some craters and a small quantity of grains on the workpiece surface with emulsion-2. Fig. 8.25(c) shows that the surface is smooth, and covered by small craters and grains. The craters on the machined surface with any emulsion show irregular shape, different size and depth, because the micro-electrical conductivity and micro-organizational structure on the SiC ceramic surface are different. Some craters are deep and connecting each other, they are produced by electrical discharges along the electrical conductive network. When emulsion-1 is used as the dielectric, part melting workpiece and tool materials can be washed away by the dielectric during ED milling, the other cools rapidly and forms grains on the

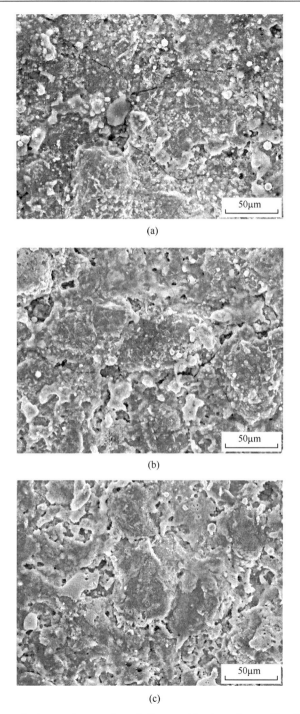

Fig. 8.25 Scanning electron micrographs of the machined surface with different emulsions. (a) Emulsion-1, (b) Emulsion-2, (c) Emulsion-3

workpiece surface. Emulsion-2 and emulsion-3 containing OP-10 have the good washing capability, and the eroded materials are washed away easily; therefore, there are less grains on the workpiece surface. The rosin is a weak acid, and it can ionize when it dissolves in the emulsion. The electrical conductivity and the acidity of the emulsion increase with an increase in rosin concentration, the electrolysis reaction will happen in emulsion-3 during ED milling, and the crater generated by a single pulse is shallow; therefore, the machined surface with emulsion-3 is smooth.

2. Compositions of the machined workpiece

During EDM, the material can be transferred between the electrodes in solid, molten, or gaseous state simultaneously [16-18]. Fig. 8.26~Fig. 8.27 show the EDS spectrum analysis of the unprocessed surface, and the machined surface with emulsion-3, respectively. The elements of the specimen are indicated by the peaks corresponding to their energy levels. It can be seen from Fig. 8.26 and Fig. 8.27 that Fe is detected on the machined surface with emulsion-3, not on the unprocessed surface. This implies that some Fe from the tool electrode diffuses into the specimen surface during ED milling. In order to determine the phases of Fe or iron compounds, the X-ray diffraction pattern of the machined surface with emulsion-3 is shown in Fig. 8.28.

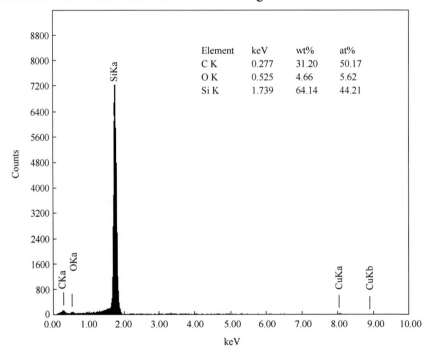

Fig. 8.26 EDS analysis of the unprocessed SiC ceramic surface

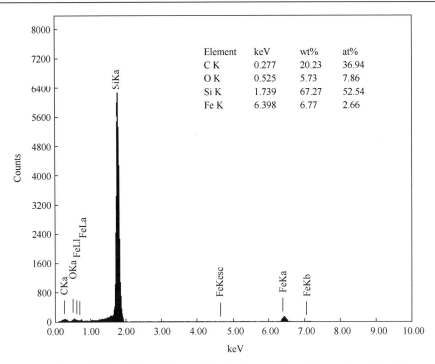

Fig. 8.27 EDS analysis of the machined SiC ceramic surface with emulsion-3

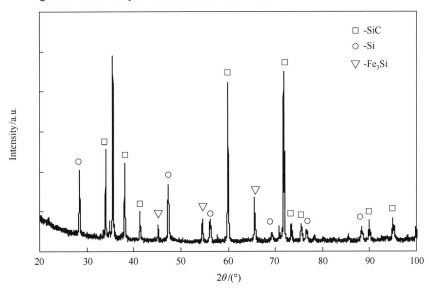

Fig. 8.28 X-ray diffraction patterns of the machined surface with emulsion-3

For comparison, the patterns of the unprocessed SiC ceramic is shown in Fig. 8.29. The peaks of Fe_3Si indicate that a modified layer forms on the machined surface, and a combination reaction takes place during ED milling of SiC ceramic, which can be described as follows:

$$3Fe + Si \rightarrow Fe_3Si$$

Because the content of O is low, there is no diffraction peak of oxygen compounds in Fig. 8.28 and Fig. 8.29.

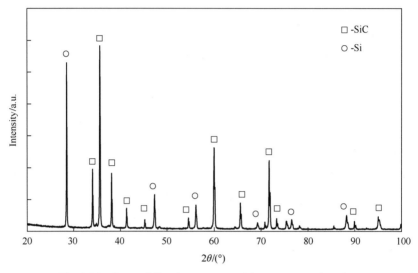

Fig. 8.29 X-ray diffraction patterns of the unprocessed surface

8.5 Conclusions

Emulsion-1 was 95wt% distilled water+5wt% emulsified oil with 25wt% ACE and 75wt% machine oil. Emulsion-2 was 95wt% distilled water+5wt% emulsified oil with 25wt% ACE, 1wt% OP-10, and 74wt% machine oil. Emulsion-3 was 95wt% distilled water+5wt% emulsified oil with 25wt% ACE, 1wt% OP-10, 2wt% rosin, and 72wt% machine oil.

Shorter pulse duration and pulse interval, higher peak voltage and peak current, and positive tool polarity are suitable for ED milling of the SiC ceramic.

In comparison with emulsion-1 and emulsion-2, emulsion-3 shows high MRR and low SR, so it shows the best machining performance.

There are many craters and grains on the machined SiC ceramic surface with emulsion-1. The quantity of craters and grains on the machined surface with emulsion-2 is small. Furthermore, the machined surface with emulsion-3 is smooth, and covered by small craters and grains.

The material from the tool can transfer to the workpiece and a combination reaction takes place during ED milling of the SiC ceramic.

References

[1] Rehbein W, Schulze H P, Mecke K, et al. Influence of selected groups of additives on breakdown in EDM sinking. Journal of Materials Processing Technology, 2004, 149(1-3): 58-64.

[2] Khan A A. Electrode wear and material removal rate during EDM of aluminum and mild steel using copper and brass electrodes. International Journal of Advanced Manufacturing Technology, 2008, 39(5/6): 482-487.

[3] Lin C T, Chow H M, Yang L D, et al. Feasibility study of micro-slit EDM machining using pure water. International Journal of Advanced Manufacturing Technology, 2007, 34(1/2): 104-110.

[4] Paulo P, Elsa H. Effect of the powder concentration and dielectric flow in the surface morphology in electrical discharge machining with powder-mixed dielectric (PMD-EDM). International Journal of Advanced Manufacturing Technology, 2008, 37(11/12): 1120-1132.

[5] Bai C Y. Effects of electrical discharge surface modification of superalloy Haynes 230 with aluminum and molybdenum on oxidation behavior. Corrosion Science, 2007, 49(10): 3889-3904.

[6] Wu K L, Yan B H, Huang F Y, et al. Improvement of surface finish on SKD steel using electro-discharge machining with aluminum and surfactant added dielectric. International Journal of Machine Tools and Manufacture, 2005, 45(10): 1195-1201.

[7] Kumar S, Singh T P. A comparative study of the performance of different EDM electrode materials in two dielectric media. Journal of the Institution of Engineers (India), Part PR: Production Engineering Division, 2007, 87(3): 3-8.

[8] Chen S L, Yan B H, Huang F Y. Influence of kerosene and distilled water as dielectrics on the electric discharge machining characteristics of Ti-6Al-4V. Journal of Materials Processing Technology, 1999, 87(1-3): 107-111.

[9] Chung D K, Kim B H, Chen C N. Micro electrical discharge milling using deionized water as a dielectric fluid. Journal of Micromechanics and Microengineering, 2007, 17(5): 867-874.

[10] Chow H M, Yang L D, Lin C T, et al. The use of SiC powder in water as dielectric for micro-slit EDM machining. Journal of Materials Processing Technology, 2008, 195(1-3): 160-170.

[11] Chow H M, Yan B H, Huang F Y. Micro slit machining using electro-discharge machining with a modified rotary disk electrode (RDE). Journal of Materials Processing Technology, 1999, 91(1-3): 161-166.

[12] Liu Y H, Ji R J, Li Q Y, et al. An experimental investigation for electric discharge milling of SiC ceramics with high electrical resistivity. Journal of Alloys and Compounds, 2009, 472(1/2):406-410.

[13] Liu Y H, Ji R J, Li Q Y, et al. Electric discharge milling of silicon carbide ceramic with high electrical resistivity. International Journal of Machine Tools and Manufacture, 2008, 48(12/13):

1504-1508.

[14] Zhao S M. The Principle, Composition, Determination and application of the surfactant. Beijing: China Petrochemical Press, 2005.

[15] Huang Q W, Jin Z H, Gao J Q, et al. Microstructure of reaction-bonded silicon carbide. Hsi-An Chiao Tung Ta Hsueh/Journal of Xi'an Jiaotong University, 2000, 34(2): 89-91, 99.

[16] Patel K M, Pandey P M, Venkateswara R P. Surface integrity and material removal mechanisms associated with the EDM of Al_2O_3 ceramic composite. International Journal of Refractory Metals and Hard Materials, 2009, 27(5): 892-899.

[17] Guu Y H, Hou M T K. Effect of machining parameters on surface textures in EDM of Fe-Mn-Al alloy. Materials Science and Engineering A, 2007, 466(1/2): 61-67.

[18] Abdullah A, Shabgard M R, Ivanov A, et al. Effect of ultrasonic-assisted EDM on the surface integrity of cemented tungsten carbide (WC-Co). International Journal of Advanced Manufacturing Technology, 2009, 41(3/4): 268-280.